Edited by
J. T. Greensmith

© THE GEOLOGISTS'
ASSOCIATION
1996

"Blind Oak Gate.
The soil of our parish consists of greensand and of chalk, and is unpropitious to earthquakes."

> E.M. Forster, "The Last of Abinger"
> 27th July, 1946

"At Abinger, as elsewhere in Surrey, a number of hollow lanes have worn down deep into the greensand. These were found useful by the smugglers, since they could move their stuff about with little chance of being seen."

> E.M. Forster, "Abinger Harvest"

"Through copse and spinney marched Bear; down open slopes of gorse and heather, over rocky beds of streams, up steep banks of sandstone into the heather again; and so at last, tired and hungry, to the Hundred Acre Wood."

> A.A. Milne, "Winnie the Pooh"

ISBN 0-900717-88-2

Early Cretaceous Environments of the Weald

CONTENTS

LIST OF FIGURES	(ii)
PREFACE	1
EVOLUTION OF THE WEALD BASIN	
Mesozoic - Cenozoic overview	2
Lower Cretaceous structure	2
STRATIGRAPHY & PALAEOENVIRONMENTS	3
ITINERARIES	
1. The Hastings Group in the core of the Weald	9
2. Coastal sections around Hastings	16
3. The Weald Clay	22
4. Folkestone	30
5. The South Downs	41
6. The Weald Anticline around Petersfield	53
7. Haslemere to Godalming	55
8. The Hog's Back	61
9. Albury	65
10. Redhill - Bletchingley - Godstone	68
11. The eastern North Downs	73
ACKNOWLEDGEMENTS	75
FURTHER READING	76

Early Cretaceous Environments of the Weald

LIST OF FIGURES

1. Formal lithostratigraphy of the Cretaceous sediments exposed in the Weald.   3
2. Schematic Wealden megacyclothem.   5
3. Chronostratigraphy of the Wealden sediments in the Weald Basin.   6
4. Itinerary locations.   8
5. Stratigraphy of the beds exposed in Philpots Quarry.   10
6. (Upper). The face at Philpots Quarry, West Hoathly. (Lower). Desiccation cracks within the Grinstead Clay Member at Philpots Quarry.   12
7. Fragments of *Equisetites*.   14
8. View looking westwards from Cliff End.   18
9. View looking eastwards of the cliffs east of Fairlight Cove, near Hastings.   20
10. Erosion of cliffs east of Fairlight Cove.   22
11. Weald Clay stratigraphy in Surrey.   25
12. Weald Clay stratigraphy in Sussex.   26
13. Section in the pit at Laybrook Brickworks.   29
14. Common fossils of the Weald Clay.   30
15. Cross-section of the cliffs and foreshore between Hythe and Sandgate.   31
16. Common fossils of the Hythe Formation at Hythe.   33
17. *Thalassinoides* burrow from the Hythe Formation, Sandgate foreshore.   34
18. The foreshore at Mill Point, Folkestone.   35
19. The 'unconformity facies' at Mill Point.   36
20. Stratigraphy of the Cretaceous succession exposed at Folkestone.   38
21. Common fossils of the Gault Clay at Folkestone.   39
22. Junction of the Lower and Upper Gault Clay at East Wear Bay, Folkestone.   40
23. Correlation of logged sections at Washington Sandpit.   46
24. Stratigraphy of the Sandgate Formation in the South Downs.   49
25. Common bivalves found in the Hythe Formation.   51
26. Correlation of the pebbly Sandgate Formation ('Bargate Beds') of the western and northern Weald.   58
27. Cartoon of the tectonic movements affecting late Hythe Formation to late Sandgate Formation deposition.   59
28. Stratigraphy of the Folkstone Formation, around Farnham.   64
29. Summary stratigraphy of the Lower Greensand exposed in the Reigate – Redhill – Bletchingly area.   69

Early Cretaceous Environments of the Weald

PREFACE

Since Kirkaldy's (1976) *Geologists' Association* Guide to the Weald, much has been published on the geology of the region. In addition, many new exposures have been created and, sadly, a great many destroyed or overgrown; such is the nature of our investigations into the geology of the Weald. Consequently, this guide can never provide a comprehensive description of the possible field trips to the many varied and temporary exposures in this populated and well-vegetated area. To avoid this work becoming out-dated too soon, we have concentrated on exposures that are likely to remain in their present state for some time. A few quarry locations are included that are of such regional geological significance that it would be unwise to neglect them at this time. Access to these areas is restricted, and recommended for specialist group visits only. In addition, quarry locations are the most likely to disappear in a short time span; thus pre-field trip reconnaissance is recommended. Our wish to document the more permanent exposures of the Weald creates a bias toward both the harder, more resistant formations (Hastings "sands"; Upper and Lower Greensands) and the coastal outcrops. To balance this, quarry locations are included in the softer formations. Although strictly in the Upper Cretaceous, we have also included the classic Glauconitic Marl exposures above the Lower Cretaceous (Albian) Upper Greensand at Eastbourne in order to maintain continuity. For those interested in the Upper Cretaceous (Upper Greensand and Chalk) we should like to draw attention to this guide's sister work by Professor R. Mortimore.

Having explained the necessary restrictions on writing this guide, we hope that the specialists will not be dismayed at the omission of their favourite abandoned clay-pit. The interested geologist will find the locations easily, observe the main features with no trouble, and be able to do this for some years to come. The specialist will find that in addition to the recommended stops, we have also listed some locations where essentially the same stratigraphy is repeated. Here, the more casual visitor may proceed to another stop or itinerary (e.g. the Ardingly Sandstone crags of Tunbridge Wells or the Pulborough Sandrock/ Folkestone Sands of the South Downs). Should our work stand the test of time as well as Kirkaldy's, especially in an age of urban population growth (where the profit made in extracting clay or sand is as great as that made from landfill!) then we shall be doing well.

Early Cretaceous Environments of the Weald

EVOLUTION OF THE WEALD BASIN

Mesozoic - Cenozoic overview

The Weald Basin covers an area of approximately 4000 square kilometres in East and West Sussex, Kent and south Surrey. The structural style and subsidence history of the Wessex Basin "complex" (which includes Weald and Channel Basins, or sub-basins of Stoneley, 1982) is broadly similar to other basins throughout the North Atlantic zone of inter- and intracontinental rifting.

An early phase of rift faulting and associated continental sedimentation in the Permian - Triassic was followed by thermal subsidence and marine transgression in the early Jurassic. A period of shallow-marine deposition in the Middle Jurassic was then succeeded by deeper waters in the late Jurassic. During latest Jurassic times, further rifting occurred, isolating the developing Purbeck and Wealden rift basins. Thermal subsidence, again associated with marine transgression, occurred in the mid-Cretaceous (Aptian to Cenomanian). Upper Cretaceous passive subsidence and Chalk deposition ended with the first stage of basin inversion at the Cretaceous -Tertiary boundary. This inversion was completed by uplift and deformation in the Miocene.

Lower Cretaceous structure

The Wealden in the type area is divided into a lower, Hastings (once termed "Beds") Group and an upper, Weald Clay Group. The Hastings Group encompasses predominantly sand- and clay-rich sediments; the Weald Clay Group comprises claystones, with minor sand and limestone beds. The subdivision of these, and younger, Cretaceous beds is summarised in Figure 1. Broadly, the present outcrop of strata is the result of Cenozoic (probably Miocene) inversion of the basin. Lower Cretaceous formations thicken towards the core of the Weald anticlinorium, away from the basin margins to the north and south. Borehole and seismic data from southern England show that the succession, at the margin of the Weald Basin, is bounded to the north by the London Platform, with a step-like series of south-throwing normal faults (Lake, 1975; Whittaker, 1985). The nature of the London Platform as a sediment source-area has been documented from sedimentary provenance studies (Allen, 1960, 1961). To the west, the basin continues in the subsurface beneath Winchester, the feather-edge possibly reaching some 50 km west to the Vale of Wardour. To the southeast, the basin continues across the English Channel to the French Boulonnais; to the southwest it is separated from the Channel Basin by the Portsdown High, or 'South Downs Swell' of Allen (1975), a contemporaneous Cretaceous structure.

Early Cretaceous Environments of the Weald

STRATIGRAPHY & PALAEOENVIRONMENTS

The Jurassic - Cretaceous boundary lies within the shallow water, lagoonal limestones, shales and evaporites of the Purbeck "Beds" of Howitt (1964), the only permanent exposure of which in the Weald Basin is in the Brightling Gypsum Mine at Robertsbridge [TQ 677218]. The change in deposition, from the marine limestones and evaporites of the Purbeck facies to the non-marine, shallow water sands and clays of the Wealden occurred in the early Cretaceous (Casey, 1963; Dodson *et al.*, 1964; later references in Allen & Wimbledon, 1991). It appears to correspond with a synchronous facies change in the marine realm which affected much of northern Europe (Rawson & Riley, 1982). The widespread nature of this led to the suggestion (Allen, 1967; 1975; 1981; 1989; Hallam, 1986) that a change from arid to humid climates was responsible. Hesselbo & Allen (1991) postulated a major unconformity just above the Purbeck - Wealden transition in the western margin of the Wessex Basin, invoking a lowering of sea-level as a significant control on sedimentation during the earliest Wealden times.

GROUP	FORMATION
Chalk (400m)	Upper Chalk
	Middle Chalk
	Lower Chalk
Upper Greensand (0-100m)	
Gault Clay (0-100m)	Upper Gault
	Lower Gault
Lower Greensand (100 - 200m)	Folkestone
	Sandgate
	Hythe
	Atherfield Clay
Weald Clay (50-750m)	Weald Clay (Upper Division)
	Weald Clay (Lower Division)
Hastings (100-350m)	Upper Tunbridge Wells Sand
	Grinstead Clay
	Lower Tunbridge Wells Sand
	Wadhurst Clay
	Ashdown
Purbeck (part)	

Figure 1.
Formal lithostratigraphy of the Cretaceous sediments exposed in the Weald. The term "beds" may be found in many older publications. Thicknesses are estimates based on recent work by the authors and publications; some may require revision.

Early Cretaceous Environments of the Weald

Two major cycles of sedimentation can be recognised within the Wealden (Hastings Group) succession (Figure 2), termed megacyclothems by Allen (1959). A third megacyclothem may be represented by the Upper Tunbridge Wells Sand and the Lower Weald Clay but knowledge is poor due to restricted exposure.

Each megacyclothem, in the lower part, consists of claystones and mudstones coarsening-upwards into a cross-bedded sandstone member, the latter reflecting a change from gently meandering to braided channel fluvial systems. Overbank deposits decrease in importance up-sequence as the channels merge into superimposed multistorey networks. The sands are capped by a thin pebble-bed sitting on a marked erosion surface. These beds contain the coarsest extra-basinal material in the Wealden succession and resulted from winnowing and strandline erosion during the regression of the Wealden mudplain. Fining upwards silty beds above these pebble beds give way to grey muds containing siderite concretions and aquatic soil horizons. These soil horizons contain rhizomes and stems of the horsetail *Equisetites* and are the remains of *Equisetites* – reedswamps which fringed the edge of the mudplain. Localised sand bodies are present within the clay formations (e.g. the Cliff End Sandstone within the Wadhurst Clay, near Hastings (see Itinerary 2) and may represent barrier bars, channel fills or small fan deltas. In general, the mudstone formations were deposited in more brackish water than the sand formations. Prominent horizons of siderite (iron carbonate) concretions formed within mudstones soon after they were buried. It is these concretions which formed the basis for the Wealden iron ore industry. From the end of the First Century B.C. until the end of the Roman occupation iron was produced in the Weald on an industrial scale. Subsequently, only scattered production took place until the end of the Fifteenth Century when the blast furnace was introduced. By the end of the next century the Wealden iron industry was the most technically advanced in Britain. A slow decline then set in and with the advent of coke-fired blasting in 1709 virtually ceased; this decline was irreversible as the Weald had no supply of coal of its own. The industry relied mainly on a few thick laterally persistent siderite horizons (e.g. the basal Wadhurst Clay Ironstone band - see Itinerary 2) and these were mainly mined from shallow bell-pits. Many of these bell-pits, although overgrown, can still be seen (numerous woods within the Weald are known as "Minepits Wood") and can be used to map the outcrop of the ironstones. The lower, arenaceous successions of the type Wealden have no known marine horizons (Topley, 1875; Allen, 1959, 1975, 1981). Allen (1959) first regarded the palaeoenvironment as being dominated by deltaic processes, controlled by remote sea-level changes. Later, in 1967, he reconsidered this hypothesis and concluded that regarding the alternating arenaceous and argillaceous cyclical deposits of the Wealden as distant products of marine regressions and transgressions was too simplistic. Massifs surrounding the Weald Basin not only provided Lower Cretaceous sediment but also, through tectonic movement,

Early Cretaceous Environments of the Weald

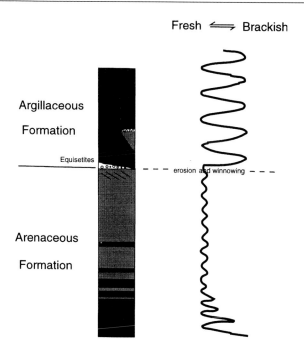

Figure 2. Schematic Wealden megacyclothem (after Allen, 1959). Dark tone = clay; light tone = sands; circles = pebbles.

ultimately controlled the stratigraphy (Allen, 1989). Poor correlation between Lower Cretaceous events in the Weald Basin and marine transgressive phases documented from the East Midlands Shelf and Paris Basin may have resulted from independent tectonic movements. Recent work by Rawson (in Cope *et al.*, 1992) suggests that low resolution in dating the non-marine facies may be a complicating factor in correlation (Figure 3). Nevertheless, the changing tectonic elevation of surrounding massifs is thought to be a fundamental control on Wealden sedimentation (Allen, 1967; 1975; 1981). The climatic change into increasing humidity resulting from these tectonic movements has been inferred by Allen (*op cit.*); Sladen (1983) and Sladen & Batten (1984). Allen (1975, 1989) and Hallam (1984, 1986) considered a climatic response to North Atlantic rifting as a fundamental stimulus to the development of the Wealden facies throughout northern Europe.

By latest Barremian times global sea-level was rising and throughout the succeeding Aptian - Albian - Cenomanian stages of the "Mid"- Cretaceous the

Early Cretaceous Environments of the Weald

effect of this increase in eustatic level was felt throughout the Weald Basin. Initially this increase in marine influence resulted in deposition of the thin marine beds found within the topmost Wealden strata (Worssam, 1978). Later, complete marine inundation of the Weald Basin took place (in the earliest

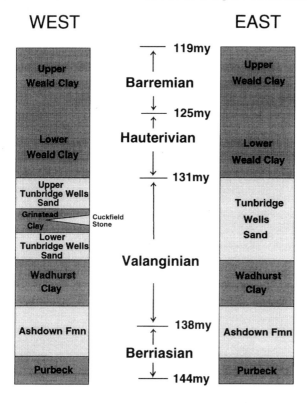

Figure 3. Chronostratigraphy of the Wealden sediments in the Weald Basin.

Aptian). From this time until the late Cretaceous marine deposition dominated, reflecting the late Cretaceous eustatic sea-level highstand. The deposition of Lower Greensand in a shallow shelf environment, the Gault and Upper Greensand in offshore (but shallow) shelf areas, and the Chalk in the European-wide sea of the Upper Cretaceous, reflects the progressive transgression of the sea across the Weald and adjoining basins.

Early Cretaceous Environments of the Weald

NATURE OF THE LOCATIONS

To aid the first-time visitor, we have constructed each part of the guide as an itinerary (Figure 4). Individual stops in the itinerary may be extracted without following the whole excursion. This may lead to minor repetition of text descriptions for the excursion follower, yet will allow those looking for a single outcrop to do so with ease.

It is essential that the guide user is in possession of the relevant 1:50,000 Ordnance Survey map for each of the areas to be visited and is *au fait* with the British National Grid reference system. The maps needed for full coverage are 179, 186, 187, 188, 197, 198 and 199.

The guide contains 37 stops and mentions another 59 additional locations. We feel this number of sites is desirable in order to provide a comprehensive number of possible places to visit in an area where building and land-fill are prevalent. These stops are often geographically widely spaced and will require additional planning, using the relevant O.S. 1:50,000 sheets, prior to your visit.

For each stop we follow this pattern:-

- how to get there (public transport, car), map
- key literature
- summary geological interest (key words)
- description (location features & diagrams)

Early Cretaceous Environments of the Weald

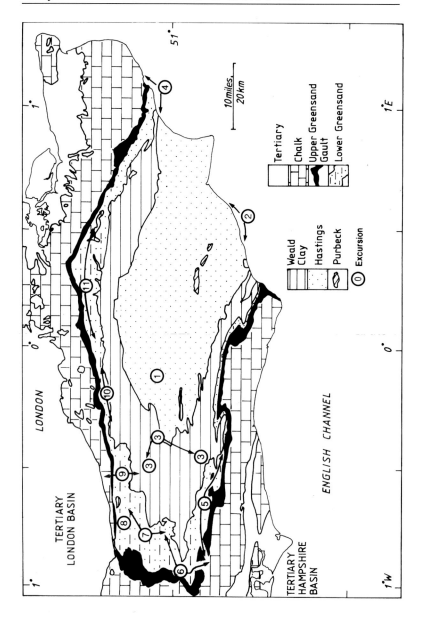

Figure 4. Itinerary locations superimposed on a map of the Geology of the Weald.

Early Cretaceous Environments of the Weald

Itinerary 1. The Hastings Group in the core of the Weald
Kevin G. Taylor

(Sheet 187, 1:50,000)

Stop 1. Philpots Quarry, West Hoathly, West Sussex, RH19 4PS [TQ 355322].

How to get there
If travelling by private vehicle park by the village church in West Hoathly. A bus service to West Hoathly village is also available from East Grinstead. From the village church walk towards the southwest, down Philpots Lane (an unpaved track) for approximately 1km and the entrance to the quarry will be found on the right. Permission to visit the quarry should be sought from the quarry manager, Mr L. Hannah, either at the pit itself or at the lodge at the top of Philpots Lane. **Please do not use hammers at this locality.**

Key Literature
Allen (1975, 1989); Taylor (1991, 1992).

Summary of Geological Interest
Junction of Lower Tunbridge Wells Sand and Grinstead Clay, Lower Tunbridge Wells Pebble Bed. Plant fossils, non-marine bivalves, siderite concretions.

Description
This quarry is the only working quarry left in the Lower Tunbridge Wells Sand and the overlying Grinstead Clay and is, therefore, the only place in the Weald where this stratigraphic interval can be well seen presently (Figure 5). The section exposes the upper part of the Lower Tunbridge Wells Sand (the Ardingly Sandstone Member), quarried here for ornamental stone, and the Lower Grinstead Clay Member (Figure 6, top). The Grinstead Clay is not utilised commercially, although in the past it has been quarried elsewhere for brick manufacture.

A large amount of detailed sedimentological work has been undertaken on the Ardingly Sandstone by Allen (1975, 1981, 1989) and much of this work is included in his major publications. Recently, work focusing more on the geochemistry of the sediments has been undertaken by Taylor (1990, 1991, 1992).

The Ardingly Sandstone Member can be divided into four units of sand, termed for quarrying purposes the Top, Lower, Rockery and Bottommost Lifts. These

Early Cretaceous Environments of the Weald

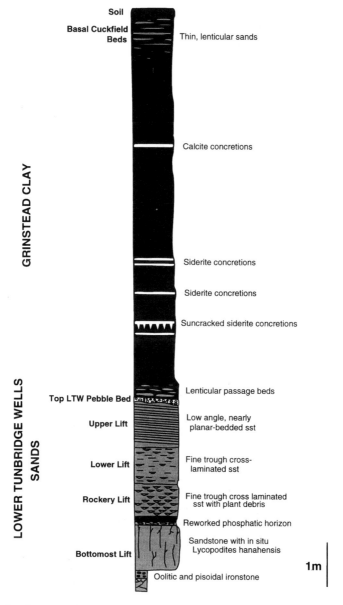

Figure 5. Stratigraphy of the beds exposed in Philpots Quarry. Lithologies as in Figure 2.

Early Cretaceous Environments of the Weald

sands have been interpreted by Allen (1975) as being deposited within the coastal margin of a fluvial braidplain. Evidence for lateritic soil-forming processes within this area is seen in the form of an oolitic ironstone at the base of the quarry (sadly no longer well exposed). This bed comprises reworked laterite ooids and pisoids whose mineralogy was altered to berthierine (an iron-rich clay mineral) upon burial. The presence of lateritic soils is in agreement with the generally proposed warm humid climate for the south of England in early Cretaceous times. In the upper parts of the Bottommost Lift the carbonaceous remains of stem of the plant *Lycopodites hanahensis* can be observed. These stems are truncated at the top of the Bottommost Lift by a horizon of reworked and oxidised phosphatic concretions. These concretions originally formed during the shallow burial of muddy sediments and were subsequently reworked by channel processes within the braidplain. Within this bed careful observation will reveal the presence of numerous fish teeth. Above the phosphatic horizon, fine sands containing predominantly low angle laminations are present. Although features within the sands are reminiscent of shoreface deposition, large scale mapping by Allen (1975) has shown that large "scoop" channel-like structures are present indicating the greater likelihood of a fluvial origin.

The Ardingly Sandstone is separated from the overlying Grinstead Clay by the Lower Tunbridge Wells Pebble Bed. This has been interpreted to be the result of braidplain winnowing during base-level rise leading to the deposition of the Grinstead Clay. Within the pebble bed a variety of clasts are found that are indicative of the geology of the London Platform to the north at the time. These include phosphate nodules, rolled ammonites and other molluscs of Jurassic origin, Carboniferous chert and silicified limestone, and pebbles of igneous and metamorphic rocks of Old Red Sandstone age. In addition teeth of reptiles (crocodiles) and fish (*Lepidotus, Hybodus*) have been found occasionally.

The quality of exposure of the overlying Grinstead Clay varies from excellent to moderately poor depending on when it was last quarried back. Thin overlying passage beds of lenticular silts, indicate quieter depositional conditions and fore-shadow the onset of a mudplain environment, represented by the dark grey shales of the Grinstead Clay. Within these passage beds fragments of the horsetail *Equisetites* can be found (Figure 7), many in growth position. They are the remnants of a fringing reedswamp that bordered the edge of the Wealden mudplain. In addition, throughout the Grinstead Clay ostracods are abundant, along with the freshwater molluscs *Neomiodon*, *Unio* and *Viviparus*. The most obvious feature of the Clay are the horizons of concretions, composed of siderite (iron carbonate), which has preferentially cemented the clay. Petrographic and geochemical work has shown that the siderite precipitated

Early Cretaceous Environments of the Weald

Figure 6a. The face at Philpots Quarry, West Hoathly. This quarry exposes the upper part (Ardingly Sandstone Member) of the Lower Tunbridge Wells Sands Formation (LTWS) and the lower part of the Grinstead Clay (GC). The boundary between these two is marked by the Lower Tunbridge Wells Pebble Bed (PB). The lower part of the Cuckfield Stone Member (CS) is present at the very top of the quarry face. (Person for scale).

Figure 6b. Desiccation cracks preserved and infilled with early siderite cement within the Grinstead Clay Member at Philpots Quarry, West Hoathly. A concretionary siderite horizon has preserved on its underside a polygonal pattern of desiccation cracks, approximately 30 cm deep. (Lens cap 5 cm across).

Early Cretaceous Environments of the Weald

from pore-water within the sediment soon after the sediment was buried (within the first few metres of burial). The concretions also continued to grow in size during later burial. The factors controlling the siting of siderite concretions are less clear. It is most likely that the original distribution of decomposing organic matter and iron oxides played a role, as well as breaks in sediment accumulation. Within some of these concretion horizons, fossil insects have recently been discovered (Andrew Ross, pers. comm.). Siderite concretions approximately 1m above the pebble bed have infilled desiccation cracks within the mudstone; (vertical siderite infills with a polygonal shape in the horizontal plane can be excavated if the Grinstead Clay exposure is good (Figure 6, bottom). The presence of such features indicates that the mudplain occasionally dried out, again suggestive of a warm early Cretaceous climate.

Stop 2. Hook Quarry [TQ 355314].

How to get there
From the top of Philpots Lane, take the road SSE for 1 km to a gate on the right near Hook Farm. Permission should be sought from Mr. L. Hannah at Philpots Quarry.

Key Literature
Allen (1960, 1975).

Summary of geological interest
Upper part of the Lower Tunbridge Wells Sand, Top Lower Tunbridge Wells Pebble Bed.

Description
This small quarry exposes a section within the Lower Tunbridge Wells Sand (Ardingly Sandstone) similar to that seen at Philpots Quarry. It was at Hook Farm, on the other side of the road, that the Top Lower Tunbridge Wells Pebble Bed was first described by Topley (1875). Allen (1960) has studied the composition of this bed in detail. It contains phosphate nodules, rolled ammonites and other molluscs of Jurassic origin, Carboniferous chert and silicified limestone, and pebbles of igneous and metamophic rocks of Old Red Sandstone (Devonian) age.

Stop 3. Sharpthorne Brickworks, Sharpthorne [TQ374329].

How to get there
Take the road from West Hoathly to Sharpthorne, turning left by the Post Office in the centre of the village. Follow the signposts to the brickworks entrance. Permission must be sought from the Quarry Manager to enter the pit.

Early Cretaceous Environments of the Weald

Figure 7. Fragments of Equisetites similar to those present within the Grinstead Clay Member at Philpots Quarry. (Scale is in cm)

Summary of Geological Interest
   Succession in the Wadhurst Clay, siderite bands.

Description
The cut face within this pit exposes the lower part of the Wadhurst Clay. This is similar in appearance to the Grinstead Clay at Philpots Quarry. Dark grey mudstones make up the majority of the succession. Prominent layers of early diagenetic siderite (clay ironstone) can be seen, and these were mined in the past as a source of iron (see Stop 1 - Philpots Quarry, for a fuller description of their formation). Some of these siderite bands are cemented around concentrations of shell material (*Unio, Neomiodon*).

Stop 4. Freshfield Lane Brickworks, Danehill, West Sussex [TQ 383264].

How to get there
   This brickpit is a little out of the way and can only realistically be visited by private transport. Take the A22 and then A275 south from East Grinstead, and at Danehill take a road heading southeast towards Freshfield. Freshfield Brickworks will be found on the right after approximately 2 km. Parking can be easily found on the grass verges outside the pit. There is a public footpath through the top part of the pit, but during working hours permission should be sought from the enquiries building at the entrance to the pit. To reach the pit itself turn right, skirting the edge of the brick kilns, for about 350 metres.

Early Cretaceous Environments of the Weald

Summary of Geological Interest
Palaeosols (ancient soil profiles) within the Lower Tunbridge Wells Sand.
Mottling of the upper part of the Wadhurst Clay.

Description
The brickpit exposes a section within the lower part of the Lower Tunbridge Wells Sand. In contrast to the stratigraphically higher section exposed at Philpots Quarry (see Stop 1 of this itinerary) the dominant lithologies at Freshfield are clay and silt. There are a number of horizons exposed that display prominent red and yellow mottling. Thin, laterally impersistent sandstones and siltstones are also present and these also commonly display mottling. These sediments were deposited on a low-energy fluvial floodplain. The thin sands represent overbank crevasse splays and the clays were deposited as interchannel floodplain deposits. The mottling of these deposits took place soon after deposition as a result of ground water fluctuation and, therefore, these horizons probably represent the traces of palaeosols (fossils soils). Groundwater table fluctuations led to changing oxidising and reducing conditions within the original soil profile and the mobilisation of iron oxides. Palaebotanical studies suggest a seasonal climate during the early Cretaceous and this may have controlled groundwater fluctuations.

In dry conditions a track leads southwards to a lower pit, which exposes the uppermost beds of the Wadhurst Clay. The junction of the Wadhurst Clay and Lower Tunbridge Wells Sand used to be exposed in this track. At this location, as is the case over most of the Weald, the top of the Wadhurst Clay is mottled red, which may also reflect soil-producing processes on the Wealden mudplain.

Stop 5. Natural Crags (Ardingly Sandstone), near Saint Hill, Stone Farm Rock. [TQ 381347].

How to get there
Take the road south from Saint Hill towards Weir Wood Reservoir. Natural exposures of the Ardingly Sandstone are present on a hill above this reservoir. Park in the car park at the sharp bend in the road [TQ 382347], cross the road and walk for approximately 0.5 km along the footpath through the woods.

Description
Here, massive sands at the top of the Lower Tunbridge Wells Sand (the Ardingly Sandstone) outcrop as large crags. These crags, in common with other natural crags of the Ardingly Sandstone, possibly formed during the Late Pleistocene in a periglacial climate. Thin units showing trough cross-stratification are overlain

Early Cretaceous Environments of the Weald

by a massive sandstone with disrupted bedding. Careful observation reveals the presence of "herring-bone" cross-strata (the name given to sedimentary structures indicating bi-directional flow). Such structures suggest tidal influence, so these sandstones may have been deposited in a tidal river mouthbar environment. Note the coarse grain size and the lack of clays, plant fossils and soil horizons indicating the high energy of current flow during the deposition of these sandstones.

The visitor interested in Ardingly Sandstone may like to note the numerous other crags present within the areas of Tunbridge Wells and Eridge Green. Many are frequented by climbers and particularly good examples include:

Eridge and Crowborough Area

Harrison's Rocks [TQ 533357]

Eridge Rocks [TQ 554357]

The Rocks [TQ 537328]

Bowle's Rocks [TQ 540330]

Tunbridge Wells area:

Toad Rock [TQ 569396]

Outcrops in the vicinity of Tunbridge Wells Common (e.g. at TQ 581400, TQ 576403, TQ 575405, TQ 576410) are notable due to the presence of intraformational coarse pebbly beds.

Itinerary 2. Coastal Sections around Hastings
Kevin G. Taylor

(Sheet 199, 1:50,000)

How to get there

Hastings town centre may be reached by the regular train service from London. Sections and coastal exposures west of Hastings (see Stop 3) and those immediately east of the town can be reached by foot. However, for the more easterly coastal exposures between Cliff End and Fairlight Cove private transport is desirable.

It is possible for the entire coastal section from Hastings to Cliff End to be walked, but a whole day must be allowed, limiting time available at the outcrop. In addition, the tides must be watched very carefully and all

Early Cretaceous Environments of the Weald

excursions should be started on a falling tide. Tide tables are generally available in Hastings from newsagents and fishing-tackle shops. Although access to and from the beach is possible at places, these can be treacherous in poor weather and cannot be relied upon.

In this guide two itineraries are suggested, one for the eastern part of the outcrop (two sections) and the other concentrating on the westerly area.

Stop 1. Eastern Coastal Outcrop - Cliff End [TQ 88901327] to Fairlight Cove [around TQ 876114].

How to get there
Drive from Hastings town centre north along the A259 to Ore. Take the road to Fairlight village and eventually Cliff End village. Park near the seafront in the village and walk west along the front to the start of the cliffs.

Key Literature
Stewart (1981), Allen (1975, 1989).

Summary of geological interest
Fluvial channel sandstones, palaeosols, bone-beds, siderite and sphaerosiderite concretions, saurian footprints, faults, fossil plants.

Description
Walking along the outcrop from the first cliff exposures you are, in effect, moving down the succession, due to a slight dip towards the NNE associated with the Fairlight Anticline.

The first cliffs encountered are in the Cliff End Sandstone (Figure 8). This comprises a series of sands and silts with a range of cross-bedding styles, and a number of channel fills. Stewart (1981) has proposed that the lower parts of this sandbody might have been deposited within a delta, whilst the upper part could represent a braided fluvial environment. Early workers placed this sandbody in the top part of the Ashdown Formation, but ostracods from the shales beneath are of Wadhurst Clay age, firmly placing the Cliff End Sandstone as a unit within the Wadhurst Clay.

Approximately 3 m above the top of the Cliff End Sandstone is the Cliff End Bone Bed. This is a lenticular body and is up to 0.2 m thick in places. However, it occurs within the upper part of the cliff and is therefore best studied from fallen blocks on the foreshore. Within this carbonate cemented coarse sandstone are pebbles of quartz, chert and reworked siderite (clay ironstone) concretions. It also contains fish scales, teeth and bones (e.g. *Lepidotus* and *Hybodus*), shark

Early Cretaceous Environments of the Weald

and reptilian remains. Early mammal teeth of *Plagiaulax* and *Loxaulux* were found by Dawson and Teilhard de Chardin (Woodward, 1911).

Just beneath the Cliff End Sandstone the basal Wadhurst Clay is represented by 1 m of grey clays with ostracods (Figure 8). Within these is a 0.1 m thick siderite concretion (clay ironstone) horizon (often termed the basal Wadhurst Clay Ironstone), which can be traced throughout most of the Weald (Worssam, 1964). It was one of the major mined horizons at the height of the medieval Wealden iron-ore industry. Petrographic studies of this and other similar siderite concretions within the Wealden sediments (e.g see Itinerary 1, Stop 1) have shown them to be composed of fine grained siderite crystals cementing the host mudstone. This siderite precipitated early (within the first few metres) during sediment burial. The abundance of reworked siderite concretions within basal lags of many channels at Hastings reflects the early nature of their formation.

Figure 8. View looking westwards from Cliff End of the succession exposed in the cliff approximately 1 km west of Pett Level, near Hastings. Within this part of the cliff the uppermost part of the Ashdown Formation (AS) can be seen. Overlying this is the Wadhurst Clay Formation (WC). The bottom part of the Wadhurst Clay Formation here is a thin succession of non-marine mudstones with siderite concretions. Overlying these is the Cliff End Sandstone (CES). This is once again overlain at the very top of the cliff by Wadhurst Clay mudstones, within which is the Cliff End Bone Bed. (The cliff in this photograph is approximately 20 m in height).

Early Cretaceous Environments of the Weald

The grey clays form a notch in the cliff from Cliff End toward Fairlight and act as a convenient marker for the Ashdown Formation - Wadhurst Clay junction. Beneath this marker are 15 m of Ashdown sands and silts which can be followed for the next 700 m to the Haddock's Reversed Fault. A number of minor channel features may be observed within these sediments; a prominent channel with an iron-rich top (possibly indicating an exposure surface) is present 400 m NE of the Haddock's Reverse Fault. The fault (Figure 9) intersects the cliff approximately 200 m NE of Haddock's Cottages [TQ 88571251] and the fault plane dips 60 degrees to the SSW. There is an approximate downthrow of 60 m on the north side, although a lack of good marker beds makes it difficult to determine the amount precisely. The low ground associated with erosion along the fault plane can provide access to the beach from the cliff top but should only be attempted in good conditions.

For the next 800 m as far as another fault, the Fairlight Cove Reverse Fault, a series of sandstones and siltstones are exposed that are approximately 60 m below the top of the Ashdown Formation. The most prominent unit is a cross-bedded unit - the Haddock Rough Unit - which extends north of the Fairlight Cove Fault for about 500 m. This unit, 10 m thick, has a highly erosional base, rising from near the bottom of the cliff at the fault to near the top of the cliff at the Haddock's Reverse Fault (Allen, 1975). Stewart (1983) has proposed that this unit was deposited as a laterally-accreting point bar deposit under conditions of variable river discharge.

The Fairlight Cove Fault cuts the cliff at TQ 88061197. This also provides access to the beach in good conditions. It has a similar trend to the Haddock's Fault, with a fault plane dipping 60 degrees to the SW and a downthrow of approximately 50 m on the N side.

SW of the Fairlight Cove Fault the sediments gradually become older until the core of the Fairlight Anticline is met [TQ 87151120]. At the headland around grid reference TQ 876114 a distinctive sandstone, the Lee Ness Sandstone, is exposed. This sandstone, up to 2 m thick, displays much small-scale lateral variability, internal erosion surfaces and load structures. The underside of the sandstone commonly contains casts of the footprints of the three-toed sauropod *Iguanodon*. The occurrence of such footprints indicates relatively shallow water conditions, probably true for many of the deposits within the Hastings Group of the Weald (Allen, 1975). The Lee Ness Sandstone may represent a lacustrine, delta or estuarine sand.

Beneath the Lee Ness Sandstone and in the core of the Fairlight Anticline 10 m of clays and silts are exposed - the oldest rocks exposed between Hastings and Cliff End. These sediments are prominently red and green mottled containing

Early Cretaceous Environments of the Weald

abundant sphaerosiderite, both as dispersed spherulites up to 2 mm across and as concretions of spherulitic siderite up to 0.5 m in size. These clays and silts were deposited as overbank and floodplain deposits and the mottling is probably a result of pedogenic (soil forming) processes. The mottling was a result of fluctuating groundwater level which led to differential mobilisation of

Figure 9. View looking eastwards of the cliffs approximately 1.5 km east of Fairlight Cove, near Hastings. Within this cliff is the Haddock's Reversed Fault (F). This fault has upthrown the Ashdown Sands Formation (AS) on the south side relative to the Wadhurst Clay Formation and the Cliff End Sandstone (CES) on the northern side.

the iron within the sediments. The origin of spherulitic siderite is less clear. The association with mottled clays suggests a pedogenic origin, an origin also suggested for Middle Jurassic spherulitic siderite (Kantorowicz, 1990).

Within this series of clays and silts (commonly termed the "Fairlight Clays") plant fossil fragments are common. Among these are ferns, cycadophytes (true cycads and Bennettitales) and conifers (e.g. *Pseudofrenelopsis*). Along the beach, boulders containing fragments of the tree fern *Tempskya* are also commonly found, although these have not been found *in situ*.

Younger sediments are seen again SW of the core of the anticline, with the Lee Ness Sandstone being particularly prominent. However, exposure is often not so

Early Cretaceous Environments of the Weald

good on this part of the section and so it is recommended that you retrace your steps back to Cliff End.

N.B. At the time of writing, work has recently been completed on coastal defences around the area of Fairlight Cove. Landslips are common in this area, causing a major threat to roads and buildings on top of the cliff (Figure 10). Time will tell what effect these defences will have on the quality of the exposure here.

Stop 2. Exposures in Old Sea Cliffs around Winchelsea.

Just north of the road through Cliff End old sea cliffs, surrounded by subsequently reclaimed marshland, can be seen at Toot Rock [TQ 892137]. Within these cliffs, a succession of upper Ashdown Formation and lower Wadhurst Clay is observed.

Similar old sea-cliffs can be traced eastwards around Pett Level, Winchelsea and Rye. Somewhat overgrown exposures of Ashdown beds are present. However, relatively good exposures of the Ashdown Formation - Wadhurst Clay junction can be observed in cliffs and lanes around Winchelsea [at TQ 90181715; 90231691; 90171642 and 90681723]. The sections show the upper sandstones of the Ashdown beds, overlain by thin Wadhurst Clay containing the basal Wadhurst siderite ironstone. On top of these is the lower part of the Cliff End Sandstone.

Stop 3. Western exposures: Cooden [TQ 70260615] to Hastings [around TQ 821095].

Along the coastline between Cooden and Hastings sporadic exposures of Tunbridge Wells Sand, Weald Clay and Wadhurst Clay occur on the foreshore as reefs and as low sea-cliffs. These exposures are generally small and degrading due to the construction of sea-defences.

A fault on the foreshore at Cooden [TQ 70260615] brings Tunbridge Wells Sand (on the western side) against Weald Clay (on the eastern side). Exposures of both occur as reefs on the foreshore. The Tunbridge Wells Sand comprises sands and silts, commonly red-mottled, with channel structures visible if exposure is good. The Weald Clay comprises grey muds with bands of siderite concretions (clay ironstone).

A cliff section at Galley Hill [TQ 759076] shows heavily faulted and disturbed Tunbridge Wells Sand consisting of mottled sands and silts, with plant debris and the occasional gastropod. Further east, at Little Galley Hill [TQ 769081] sands and silts occur of the stratigraphically lower Ashdown Formation.

Early Cretaceous Environments of the Weald

Figure 10. Erosion of cliffs 0.5 km east of Fairlight Cove, near Hastings. This clearly shows the impact that such erosion has upon properties close to the cliffs. Recently, coastal defences were built in this area to minimise such impacts.

Near St Leonards, patches of sand from within the Wadhurst Clay outcrop as reefs at Goat Ledge [TQ 802087]. These sands are known as the 'Tilgate Stone', a calcareous siltstone which occurs within the lower part of the Wadhurst Clay. A fault [TQ 803089] terminates the 'Tilgate Stone', bringing down Tunbridge Wells Sand onto the foreshore.

Crags around Hastings Castle [TQ 821095] expose massive sandstones from within the top part of the Ashdown Formation.

Itinerary 3. The Weald Clay of the Weald
Andrew Ross

(Sheets 187, 197, 1:50,000)

The Weald Clay consists of a thick succession of clays forming a low-lying C-shaped outcrop in the Weald. It was laid down as a non-marine flood-plain deposit in the subsiding Wealden Basin with occasional marine incursions from the Boreal Sea to the north (Rawson, 1992, In: Cope *et al.*).

Early Cretaceous Environments of the Weald

Lithologically it consists of grey silty clays, dark grey shaley clays, brown clays, red and green mottled clays, with beds of sandstone, freshwater limestone and ironstone. The red clays are often associated with the many sandstone beds that occur and indicate that subaerial oxidisation has taken place.

Stratigraphically the Weald Clay Group has been divided into lower and upper formations which correspond roughly to the Hauterivian (Hau) and Barremian (Brm) stages respectively (Worssam, 1978). The sandstone and limestone beds were numbered by the British Geological Survey and were grouped into larger units: 1 consists of the Horsham Stone, which is the only sandstone unit in the Weald Clay that has been used as a building stone; 2 consists of beds of freshwater Small *'Paludina'* and *'Cyrena'* Limestones separated by clay; 3, 5 and 7 are sequences of alternating sandstone and clay beds; 4 and 6 are Large *'Paludina'* Limestones, which were also used locally for building stone, and 8-11 is a sequence of Large *'Paludina'* Limestone, sandstone and clay beds. The base of sandstone Bed 3a is the boundary between the Lower and Upper Weald Clay and is also the Hauterivian/Barremian boundary in the area northwest of Horsham. This boundary has often been misplaced within unit 5 in the past due to a confusion between BGS and Topley (1875) bed numbers (Ross & Cook, 1995). All the units vary in thickness laterally and some beds only occur locally. Unit 7 forms a thick sequence in the western part of the outcrop whereas unit 8-11 only outcrops in the northern part of the outcrop where 7 peters out. When the maximum thicknesses of the units are added up, the total thickness for the Weald Clay exceeds 700 m.

Palaeontologically the Weald Clay displays a fauna and flora typical of a non-marine deposit consisting of fossils from two different environments. The aquatic environment supported a fauna of fresh to brackish bivalves, gastropods, ostracods, conchostracans, isopods, fish and sharks, whereas the terrestrial environment supported dinosaurs, pterosaurs, ferns, cycads, conifers and a diverse insect fauna. Crocodiles and turtles lived in both environments.

The Weald Clay is only exposed inland in brickpits, stream gullies and freshly dug temporary exposures such as roadworks or ditches. The best exposures are in the working pits in Sussex and Surrey, as once an exposure is no longer dug, the clay soon weathers and becomes vegetated. There are currently 6 working pits in Surrey and 6 in Sussex. The stratigraphy of the Weald Clay of Surrey and Sussex and the position of each pit is shown in Figures 11 & 12 (L. W. C. = Lower Weald Clay, U. W. C. = Upper Weald Clay, thicknesses of numbered beds exaggerated in most cases). The pits vary considerably in size from the tiny Ashpark Brickworks through to the large workings of Clockhouse Brickworks. Most allow organised trips by geological societies but only a few allow visits by individuals. Prior permission is needed and addresses of the

Early Cretaceous Environments of the Weald

works are given below. To visit all three in a day would be difficult as they are all some distance apart. It would be better to pick two out of the three for a good days field work. A trip to Laybrook Brickworks could also be combined with a visit to Warminghurst Church (Stop 2; Itinerary 5 to the South Downs) which is built of Large *'Paludina'* Limestone. The pits are extremely variable in the numbers and types of fossils that can be found on a day depending on whether a fossiliferous part of the pit has been dug recently or has been washed by rain. If the ground is wet take extra care as it will be extremely muddy and the pits are potentially dangerous. Do not go anywhere near the ponds in the bottom of the pits and avoid visiting in the winter months. Some trips will yield little, others much more, so one trip a year is recommended to build up a representative collection.

Stop 1. Clockhouse Brickworks [TQ 173386].
Permission to visit obtainable from: Works Manager,
Butterley Brick Limited, Clockhouse Brickworks, Horsham Road,
Capel, Nr. Dorking, Surrey RH5 5JL
(Sheet 187, 1: 50,000)

How to get there
By car. The works are situated on the east side of the A24 just north of the Surrey/Sussex border.

Key literature
Worssam (1978), Horne (1988), Jarzembowski (1991), Gallois & Worssam (1993).

Summary of geological interest
Lower Weald Clay; BGS Bed 3 (Clockhouse Sandstone); gastropods; ostracods; conchostracans; insects.

Description
Detailed sections are given in both Worssam (1978) and Horne (1988). The lithology here consists predominantly of grey laminated silty or shaley clay with thin siltstone horizons. A few of the siltstone horizons are packed with conchostracans and at other horizons freshwater ostracods (mainly *Cypridea*) are abundant enough to form thin (few mm thick) limestones. The lowest recorded beds consist of dark grey/green shaley clay containing abundant specimens of the brackish-water gastropod *Paraglauconia shipbornensis* and various species of bivalve. This is known as the *'Cassiope'* Band and indicates a marine incursion from the Boreal Sea to the north. Unfortunately this band is not now exposed because the deepest part of the pit was recently filled with overburden and the current workings do not go deep enough. Towards the top of the pit occurs a 0.5 m thick, micaceous, ripple-bedded and

Early Cretaceous Environments of the Weald

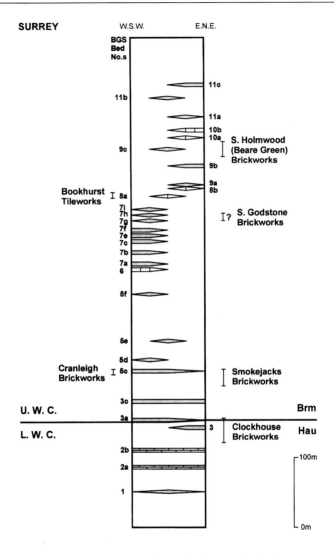

Figure 11. *Weald Clay stratigraphy in Surrey, indicating the horizons exposed in the working pits. U.W.C & L.W.C. = Upper & Lower Weald Clay. Hau = Hauterivian, Brm = Barremian.*

Early Cretaceous Environments of the Weald

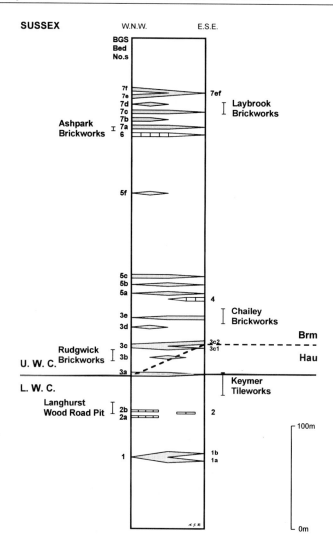

Figure 12. Weald Clay stratigraphy in Sussex indicating the position of working pits.

Early Cretaceous Environments of the Weald

suncracked sandstone, which forms BGS Bed 3 and is known as the Clockhouse Sandstone (Figure 11). It contains 1 cm wide burrows which were originally identified as the marine indicator *Ophiomorpha*, but are now recognised as the fluvial trace fossil *Beaconites*. These could just have easily been formed in freshwater conditions. The clay contains large siltstone (or fine sandstone) basin casts and lenses which have yielded an important insect fauna and rare isopods. The most common insect remains are beetle wingcases which look like rounded brown blobs. Cockroach, bug and fly wings are fairly common, but dragonfly, grasshopper, termite, scorpionfly, lacewing, wasp and caddisfly wings are much rarer. The square or rhomboidal scales and round teeth of the fish *Lepidotes* and teeth and pieces of fin spine of the freshwater shark *Hybodus* can occasionally be found in the clays. Also in the clays can be found crushed specimens of the freshwater gastropods *Viviparus fluviorum* (Large *'Paludina'*) and *V. infracretacicus* (Small *'Paludina'*). Pieces of the fern *Weichselia* and horsetail *Equisetum* have also been recorded from here. Even a vertebra of the dinosaur *Iguanodon* has been found here. Some common fossils are shown in Figure 13.

Stop 2. Bookhurst Tileworks [TQ 076395].
Permission to visit obtainable from: Mr R. Swallows, Swallow's Tiles (Cranleigh) Ltd, Bookhurst Brick and Tile Works, Cranleigh, Nr. Guildford, Surrey GU6 7DP (Sheet 187, 1:50,000)

How to get there
By car. The works are situated on the south side of the B2127 about halfway between Cranleigh and Ewhurst.

Key literature
Thurrell, Worssam & Edmonds (1968).

Summary of geological interest
Upper Weald Clay; BGS Bed 8a (Large *'Paludina'* Limestone); gastropods; bivalves; ostracods, insects.

Description
A detailed 14 m section is given on p. 56 of Thurrell *et al.* (1968), although only the top 8 m is currently being worked. The lithology consists predominantly of grey and brown shaley clay with beds of Large *'Paludina'* Limestone, siltstone and ironstone. This is the only working pit that exposes Large *'Paludina'* Limestone, which is numbered BGS Bed 8a. The limestone is packed with well-preserved specimens of the gastropod *Viviparus fluviorum* (Figure 13) and is exposed in the base of the pit in the southern corner. Individual specimens of *Viviparus* can be picked up loose on the surface; they

Early Cretaceous Environments of the Weald

have been weathered out of loose blocks of the limestone. The scales of *Lepidotes* can occasionally be found in the limestone. Ostracods have been recorded from most of the beds and the freshwater bivalve *Unio* is also present. The siltstone (bed 10 in Thurrell *et al.* section) has yielded an insect fauna consisting mainly of beetle wing-cases, with a few cockroach, bug and fly wings, and even a wasp wing.

Stop 3. Laybrook Brickworks [TQ 115188].

Permission to visit obtainable from: Mr I. R. Lissamore (Factory Manager), Ibstock Brick Laybrook, Goose Green Lane, Thakeham, Nr. Pulborough, W. Sussex RH20 2LW
(Sheet 197, 1:50,000)

How to get there
By car. The works are situated on the northeast side of the B2133, just southeast of the junction with the B2139.

Key literature
Young & Lake (1988).

Summary geological interest
Upper Weald Clay; fish, sharks, conifer twigs.

Description
This site (formerly Goose Green Brickworks) lies near the top of the Weald Clay and is roughly equivalent to unit 7, but there are no sandstone beds in this area (Figure 12). A section is given on p. 20 of Young & Lake (1988), but this differs from what can be seen today. The pit exposes about 10 m of clay and silt and a section through the unweathered sediments in the bottom of the pit (measured in 1995) is shown in Figure 13.

There is a distinctive ironstone bed in the base of the pit with trace fossils on its upper surface which are either burrows or root casts. Thicker pieces of the ironstone, and the clay immediately below it, contain twigs (up to 1 cm wide) of the conifer *Pseudofrenelopsis* which superficially resemble the stems of horsetails. The thicker pieces of ironstone have also yielded a few poorly preserved beetle wing-cases and cockroach wings. Fish remains are common in the grey, silty clays above the ironstone and can be picked up from the surface of the face after rain. They include *Lepidotes* scales and teeth, *Hybodus* teeth, fin-spines and cephalic spines (hooked spines from behind the eyes of male sharks), pieces of jaw and teeth of other types of fish and fish coprolites. A crocodile and pterosaur tooth have also been found here. The grey shaley clays also contain ostracods, most of which belong to the genus *Cypridea* (Figure 14) including *C. spinigera* which has a single, large,

Early Cretaceous Environments of the Weald

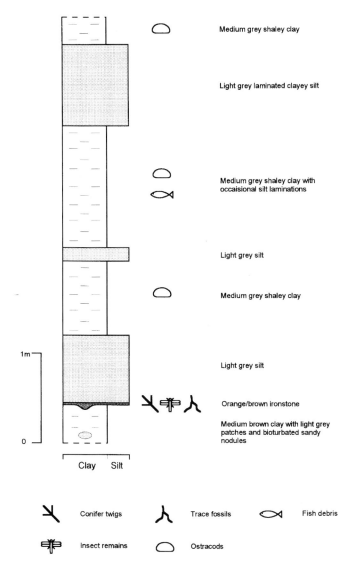

Figure 13. Section measured in 1995 at the base of the pit in Laybrook Brickworks.

Early Cretaceous Environments of the Weald

characteristic spine sticking out of the side of each valve. A new pit was started in 1991, 200 m further east from the above pit. This new pit has not yielded any fossils as yet but hopefully will in the future. A borehole from this

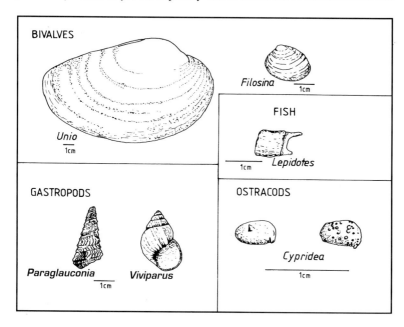

Figure 14. Common fossils of the Weald Clay (all locations in Itinerary 3). Note the variable length of the scale bar - Unio is <u>much</u> larger than ostracods!

new site reached Large *'Paludina'* Limestone (BGS Bed 6) at a depth of exactly 30 m from the top of the soil.

Itinerary 4. Folkestone
Alastair Ruffell

(Sheet 179, 1:50,000)

How to get there
>Folkestone is eminently accessible by all forms of transport. The first three stops are each walkable from the main railway station, whilst the Chalk exposures and the whole excursion are best visited using private transport. The railway station links with Dover and London (Charing Cross, Waterloo, London Bridge).

Early Cretaceous Environments of the Weald

Stop 1. Sandgate Foreshore, Sandgate [TR 194348 to TR 199349].

How to get there
   At any point on the A259 between the two grid references, exposures of the Hythe Formation can be seen on the foreshore beyond the sea wall. These are only worth visiting at low tide: tide-tables for the area can be purchased at fishing-tackle shops and newsagents in the area. Quality newspapers give tide conversions from London Bridge for Folkestone Harbour (as an estimate). Be warned, storms can leave a great deal of shingle across the Sandgate foreshore and the exposures are always partly covered!

Key literature
   Casey (1961); Smart, Bisson & Worssam (1966); Ruffell (1992).

Summary geological interest
   Rotational slumps; rag and hassock; ammonites; Hythe Formation type location.

Description
The visitor to Sandgate foreshore is struck by the variable (occasionally vertically-dipping) orientation of the well-bedded Hythe Formation. The high dip is a product of coastal landslipping and the slope failure of rock strata as rotational fault-blocks (Figure 15).

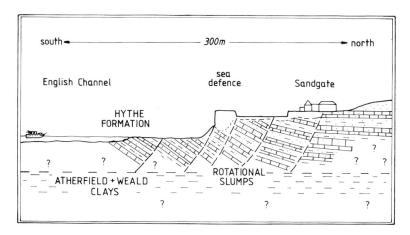

Figure 15. Cross-section (a few hundred metres across) of the cliffs and foreshore between Hythe and Sandgate showing the rotational fault-blocks resulting from high groundwater pressure in the Hythe Formation above the impermeable Atherfield and Weald Clay.

Early Cretaceous Environments of the Weald

The Hythe Formation, between Hythe itself and Sandgate, comprises 10 - 50 cm thick alternating layers of sandy, blue/grey limestone (rag or ragstone) and clay-rich, brown/green sand with glauconite (hassock). The proportions of different minerals within each rag or hassock bed may vary widely, some distinctive hassock layers being almost black fossiliferous clays, others being (effectively) uncemented ragstone. The Hythe Formation in this area is 20 m to 30 m thick; accurate measurements are hampered by stratigraphic repetition through slumping and by erosion at the Hythe - Sandgate Formation boundary (Stop 2, Mill Point). The Hythe Formation around Sandgate rests on the Atherfield Clay Formation, which in turn rests on the Weald Clay Group. Marine erosion of these soft formations at the "toe" of each rotational fault-block, coupled with the pressure of groundwater within the Hythe Formation, creates the landslip. Secondary rotational failure occurs within the Hythe Formation, along the junction between the hassock and ragstone. This has the same origin as the larger slips, a result being a repetition of the beds. The origin of the alternating rag and hassock beds is discussed by Casey (1961, p.520), who concludes that calcite cementation of the ragstone was early (sometime just after deposition and burial). It appears likely that this early diagenetic precipitation of calcite occurred (preferentially) along coarser layers in the bedding.

Weathered surfaces of ragstone on the foreshore at Sandgate expose a wide variety of fossils. These are usually encrusted with present-day marine flora and fauna, including the mussels collected in this area by commercial traders: for this reason the visitor should avoid hammering unnecessarily as the specimen is unlikely to be extracted in any better condition than you observe it in the rock. The two most common fossil types are (large) bivalves and ammonites. More scarce are brachiopods, gastropods and belemnites. The bivalves include the oyster *Aetostreon* and the trigonid *Linotrigonia*, as well as rare *Gervillella* and *Yaadia* (Figure 16). Ammonites include the large coiled *Tropaeum* and *Cheloniceras*, rare *Dufrenoyia* and the uncoiled (heteromorph) *Australiceras*. Brachiopods, where found, occur in clumps or 'nests' (Middlemiss, 1962a) of *Rhynchonella* or *Terebratula*. Within the undeformed hassock layers the belemnite *Neohibolites* may be found. The fossils form a distinctive Lower Aptian fauna, the correlation and significance of which is discussed by Casey (1961). Also present are large (up to a metre diameter) radiating networks of the trace fossil *Thalassinoides* (Figure 17).

Stop 2. Mill Point, Sandgate Toll Road (Saxon Shore Way) [TR 216352 to TR 221353].

How to get there

Between Sandgate and Folkestone the lower shore road (Saxon Shore Way) is a wooded lane adjacent to the coast. For half of its length this is a

Early Cretaceous Environments of the Weald

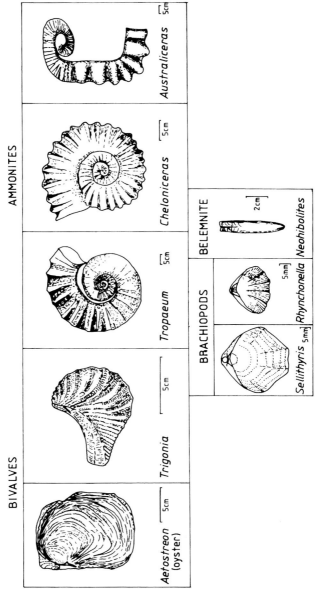

Figure 16. Common fossils of the Hythe Formation (at Hythe) found on weathered wave-cut platform bedding surfaces.

toll-road. The exposure at Mill Point is marked as rocky outcrop on the 1:50,000 sheet, between the Saxon Shore Way and Toll. The shore may also be approached from Folkestone town centre *via* one of the zig-zag paths from cliff to shore. The exposure at Mill Point itself comprises the hardened upper beds of the Hythe Formation, overlain by the softer Sandgate Formation. The exposures are almost all on the shore (some Sandgate Formation sands can be seen in a small hillock north of Mill Point) and subject to both daily (tidal) marine and seasonal inundation by water and shingle! The objective of the visit, the exposures at the east of Mill Point, are visible from the road and adjacent walkways: a small zig-zag path through terraced beach-huts brings the visitor to Mill Point.

Figure 17. Thalassinoides burrow from the Hythe Formation, Sandgate foreshore. The hammer is about 30cm long.

Key Literature
Casey (1961); Hesselbo *et al.* (1990); Ruffell (1990).

Summary geological interest
Hythe - Sandgate Formation junction and associated exotic conglomerate or 'unconformity facies'.

Description
Walking east from the zig-zag path toward the easternmost exposure of Mill Point, rotational slump-blocks, characterised by variable and sometimes high

Early Cretaceous Environments of the Weald

Figure 18. Photograph taken facing south of the foreshore at Mill Point, Folkestone. The centre mass of rock on the wave-cut platform comprises Hythe Formation (a hardened "Rag" bed); the 'unconformity facies' developed at the Hythe - Sandgate junction may be found around the concrete groyne to the left of the picture.

dips, can be seen in the Hythe Formation at low tide. At the rock outcrops furthest east (from where Folkestone Harbour should be visible), the grey weathering and mostly limpet, barnacle and mussel encrusted rocks of the Hythe Formation (Figure 18) give way to a red-weathering conglomerate with a sand matrix of the basal Sandgate Formation. The visitor should not be misled by appearances here for the conglomerate looks similar to beachrock. This is compounded by the soft Sandgate Formation being removed by erosion, to be replaced by loose flint beach cobbles. This conglomerate is what Mill Point is famous for: the beds developed at the Hythe - Sandgate Formation boundary. The conglomerate was described by Hesselbo *et al.* (1990) as 'unconformity facies', having formed during a period of erosion and reworking. Casey records that at this location, the equivalent of about one Aptian ammonite zone is absent from the Sandgate - Hythe junction. During this time period, many sedimentary rock-types were deposited, only to be eroded away, leaving a few pebbles in the conglomerate as an indication of their former existence. Throughout such reworking (of which there must have been many phases) there occurred growth of phosphate and carbonate minerals: these form nodules that are also incorporated into the conglomerate, the pebbles of which show occasional

Early Cretaceous Environments of the Weald

inhabitation by contemporaneous rock-boring organisms. These borings should not be confused with similar borings at the present day by sponges and shipworms. The Aptian colonisation is identifiable by the infilling of Sandgate Formation matrix, and by associated mineralisation (Figure 19).

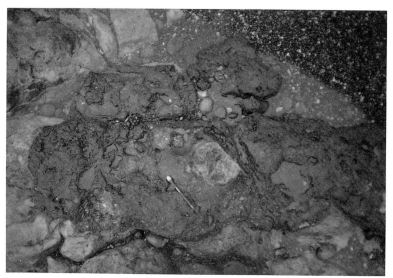

Figure 19. The 'unconformity facies' of Mill Point, Folkestone. The left and bottom margins of the photograph show Hythe Formation, the pencil (5 cm long) rests on dark Sandgate Formation with angular blocks of rafted Hythe Formation within it (to the right of the pencil).

The rock exposure at Mill Point does not appear spectacular, yet the location deserves detailed examination, if only to envisage the processes of sedimentation and habitation on the boulder-strewn sea-floor of this margin of the Weald Basin over 100 million years ago. Thus in one location, Mill Point affords the geologist an opportunity to observe many varied sedimentary rock types. Fossils from the conglomerate bed are rare, and any that are found should be taken to a museum, or shown to an expert. The fauna of the Hythe Formation, described in Stop 1 (Sandgate) may also be found in the rotated fault blocks to the west and south of the Hythe - Sandgate junction.

Stop 3. Copt Point, Folkestone Harbour [TR 238363 to TR 245368].

How to get there

Copt Point may be approached from the promenade leading from Folkestone Harbour (and railway terminus), where there is limited

Early Cretaceous Environments of the Weald

parking, or from the road to the north of Copt Point itself, where there is roadside parking.

Key literature
Casey (1961); Gale (1989); Owen (1976); Owen (1988); Hesselbo *et al.* (1990); Ruffell (1990).

Summary geological interest
Type location for the Folkestone Formation and junction with the Gault; phosphate nodule beds; abundant fossils.

Description
Along the promenade and sea-defence wall of Folkestone Harbour, the silica and carbonate cemented Folkestone Formation sands dip gently east at 2° - 5°. The low cliffs are generally inaccessible for close examination. At the eastern end of the promenade the concrete wall gives way to slipped Gault Clay on Folkestone Formation sands. Here, the complex series of nodule beds at the Folkestone Formation - Gault junction can be examined. The most easily identified horizon is the Sulphur Band, as this double row of black phosphate nodules occurs at the junction between yellow/white Folkestone sands and black sands of the basal Gault. From this horizon one may examine the Main *Mammillatum* bed some 80 cm below, or the *dentatus* nodule bed 50 cm - 60 cm above (Figure 20). Each nodule horizon shows different petrographic features, examinable upon cracking open with a hammer. The most notable non-examinable feature is the high level of radioactivity associated with the Sulphur Band nodules (Ruffell, 1990).

The passage north from Copt Point into East Wear Bay can be made on the slipped Gault Clay below Folkestone Warren: only high tides preclude this. Occasional exposures of the lower beds of the Gault can be found on the slipped slopes north of Copt Point, and after storms in East Wear Bay. All exposures provide abundant fossil material (Figure 21), although most notable are the erosive horizons of the *cristatum* nodule bed and *auritus* nodule bed (Figure 20). The former separates Lower from Upper Gault and marks the boundary between dark clays below and lighter clays above (see Figure 22 and Owen, 1971). As in the Folkestone - Gault Formation(s) boundary, these non-depositional episodes are marked by a phosphatic nodule bed. Ammonites and small bivalves are found both rolled, overgrown with phosphate, and intact alongside the nodules. Owen has demonstrated that this erosive horizon can be correlated throughout the Weald Basin, and is thus of regional significance. Without good exposure of the whole succession, or some experience of the lithology and ammonite biostratigraphy, isolated intra-Gault nodules are impossible to tell apart.
The upper beds of the Gault at Folkestone are the stratigraphic (time)

Early Cretaceous Environments of the Weald

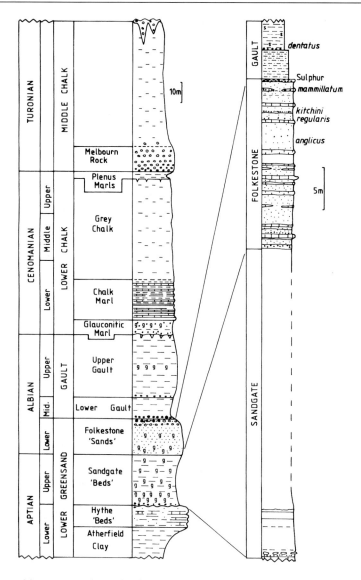

Figure 20. Stratigraphy of the Cretaceous succession exposed at Folkestone with details of the Albian succession (Folkestone sands - Gault claystone transition) exposed at Copt Point. Partly re-drawn after unpublished work by J. Hancock. Most of the Sandgate Formation is unexposed around Folkestone.

Early Cretaceous Environments of the Weald

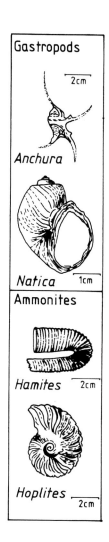

Figure 21. Common fossils of the Gault Clay, usually found washed out by rain and tides between Copt Point and East Wear Bay.

equivalents of the Upper Greensand exposed in the western Weald. The beds are intermittently exposed in the northeast of East Wear Bay. Usually only isolated exposures of the light - dark bedding rhythms of the Upper Gault can be observed. Gale (1989) describes Price's (1874) Bed XII as a useful marker horizon close to the top of the Gault succession and comprising a dark grey-green silty clay. A second marker bed, Price's Bed XI, is not so commonly exposed. Gale describes this bed as "sparsely fossiliferous; it contains many light buff coloured phosphatic nodules". Similar nodules are found in Bed XII above, although here they have been bored and encrusted like those of the Sulphur Band at Copt Point.

Figure 22. Junction of the Upper and Lower Gault Clay at East Wear Bay, Folkestone. The cristatum nodule bed may be observed just below the lithological boundary.

Early Cretaceous Environments of the Weald

Itinerary 5. South Downs
Alastair Ruffell & from Bristow & Morter (1983); Young (1978)

(Sheets 197, 198, 199 1:50,000)

Stop. 1 Cow Gap, Eastbourne [TV 59559563 to 59809580] (Sheet 199).

How to get there
Eastbourne is well-served by road and rail, with easy connections to London. The Cow Gap section however is 3.5 km to 4 km southwest of Eastbourne town, the walk taking one across steep chalk cliffs approaching Beachy Head (the 'end' of the South Downs). Access by private transport is easier: the driver may park at the Beachy Head pub [TV 5887 9582], opposite the coastguard station, or on The Pound Road [TV 6020 9684]. The sections to be studied are along the landslips on the eastern cliff of Beachy Head. The path recommended is to the east of the South Downs Way, along the landslips. Pathways here are stable underfoot, but not recommended in high winds.

Key literature
Kennedy (1967); Young (1978).

Summary geological interest
The basal Chalk succession at Eastbourne provides an excellent exposure of the Upper Greensand and Chalk, yet this area of classic geology is poorly documented. The most user-friendly guide to the area is unfortunately difficult to obtain (Young, 1978), and so the following extracts are taken directly from this work, with permission of the author.

All from Young (1978):
Outcrop geology
"The coast between Eastbourne and Beachy Head provides fine sections of beds ranging in age from Upper Albian to Coniacian (Upper Cretaceous). The object of this excursion is to examine the highest Albian and lowest Cenomanian sediments exposed at Cow Gap. At Eastbourne these stages are represented by Upper Gault, Upper Greensand and Lower Chalk and (Eastbourne) provides the only complete section through the Upper Greensand in this part of southern England".

"Widespread deposition of the Chalk followed the Cenomanian transgression across much of central and western Europe. Throughout the whole of south-east England the basal bed of the Lower Chalk consists of a glauconite-rich, sandy marly limestone known as the Glauconite Marl (incorrectly termed 'Chloritic Marl' in older literature such as Jukes-Browne & Hill, 1990). This distinctive bed passes up into rhythmically bedded hard chalks and soft marly chalks as described by Kennedy (1967)".

"The Section at Cow Gap"
"The Cow gap section is approached by the most easterly footpath from clifftop across landslipped chalk cliffs. Rotational slips have developed here in post-Pleistocene times by slipping of the Chalk and Upper Greensand on the underlying Gault. From the footpath the complex pattern of slipped masses can be seen clearly in the wave-cut platform at low tide. The section [recommended] to be visited consists of a relatively undisturbed block of sediments within the landslips."

"The topmost beds of the Gault are intermittently exposed on the beach and exhibit a burrowed contact with the lowest bed of the overlying Upper Greensand which consists here of a dark green glauconitic sand. The precise position of the topmost Gault within the Upper Albian [ammonite] zonal scheme is uncertain here; the nearest diagnostic macro-fossil recorded from this locality is a fragment of *Mortoniceras* cf *inflatum* (J. Sowerby) found at approximately 7 m below the base of the Upper Greensand and indicative of the *inflatum* zone (Kennedy, 1967). Clearly, the base of the Upper Greensand here represents an important sedimentological change."

"The basal glauconite sand of the Upper Greensand passes up into very finegrained, slightly micaceous calcareous sandstones in which [there] occur several well-defined bands of hard calcareous doggers. These sandstones are well-exposed both in the base of the low cliffs and on the beach and form the prominent reef known as Head Ledge seen at low tide immediately to the south". "Macrofossils are extremely rare ... and include the following: from the Upper Greensand at Eastbourne:- '*Nautilus* sp., *Kingena lima* (? syn *K. spinulosa* (Davidson & Morris), *Pecten orbicularis* (syn *Entolium orbiculare* (J.Sowerby), *Plicatula pectinoides* (syn *P. gurgitis* Pictet & Roux), *Holaster laevis* (Deluc) and *Jerea* sp'. Though generally regarded as being of Upper Albian age," ......"much of the Eastbourne Upper Greensand may be of early Cenomanian age."

"The contact [of the Upper Greensand] with the [overlying] Glauconitic Marl is not easily distinguished though on close inspection it can be seen as an intensely burrowed surface with numerous crustacean burrows, including the type referred to as *Thalassinoides* sp. (Kennedy, 1975), that are up to 3 cm in diameter and penetrate as far as 0.65 m into the underlying [greensand] sediment. Phosphatic nodules are common within the Glauconitic Marl and include phosphatised fossil fragments, fossil moulds and concretions some of which may have been derived from the [underlying] Upper Greensand. Well-

Early Cretaceous Environments of the Weald

preserved sponges are common here in the Glauconitic Marl which can be seen to pass upwards into the characteristic rhythmically bedded Lower Chalk in the adjacent cliffs."

"The South Downs end abruptly in the imposing vertical Chalk cliffs at Beachy Head which are some 160 m high and the highest cliffs along the south coast."

Stop 2. Hythe Formation, Warmingshurst [TQ 11701675] (Sheet 198).

How to get there
Warmingshurst Church is sited on a cross-roads, with Thakenham 1.5 km to the west, Rock and Washington 3.5 km - 4 km to the south and Ashington 1.5 km to the east. The visitor with no private transport could only reach Warmingshurst by taking the bus to Ashington and walking west. The Hythe Formation is exposed in the road cutting to the south of the church. Access is open (public road), nonetheless the cutting is used as a thoroughfare for cattle and so no rock debris should be left lying on road or verge. Hammering must be avoided. This location is 3.5 km (by road) south of the Weald Clay exposed in Laybrook Brickpit (Itinerary 3, Stop 3).

Key literature
Reid (1903); White (1924); Ruffell (1992).

Summary of geological interest
Most easterly outcrop of the Hythe Formation (South Downs); early diagenesis and coarsening-up cyclicity.

Description
The well-bedded Hythe Formation exposed at Warmingshurst was analysed by Ruffell who suggested that the bedding is a consequence of primary grain-size variation, enhanced by diagenetic growth of ferroan calcite in the coarse horizons. Such coarse beds cap upward-coarsening cycles, found throughout the Hythe Formation 'rag and hassock' of the South Downs. The change in grain size is not always due to significantly larger fragments being introduced into the depositional system, but to the gradual upward depletion (through winnowing) of clay matrix, thus increasing the mean grain-size. At Warmingshurst, the top of each coarsening-up cycle has few associated fossils; when found these are commonly large oysters. Such large bivalves indicate firm seafloor conditions when originally deposited. Other Hythe Formation outcrops in the area show the development of shell-rich layers. In contrast, at Warmingshurst, the fauna is found within hassock beds. Ruffell listed the bivalves *Pseudaphrodina*, *Pterotrigonia*, *Yaadia* and *Plicatula*. From

Early Cretaceous Environments of the Weald

the mapped outcrops of Atherfield Formation (clay) to the south, this outcrop is thought to be stratigraphically low in the Hythe Formation succession.

Stop 3. West Chiltington & Nutbourne [TQ 087183, TQ 083183, TQ 083190, TQ 073192] (Sheet 197).

How to get there
    North of West Chiltington (2 km) and Nutbourne (0.5 km) are a number of "sunken" lanes with cuttings exposing Hythe and Sandgate Formation sandstones. Access by public transport is difficult, although the strong walker can approach from the train station at Pulborough, some 3 to 4 km to the west (see Pulborough stops).

Key literature
    Bristow & Wyatt (1983); Ruffell (1992).

Summary of geological interest
    Hythe - Sandgate Formation junction, smectite clays.

Description
The Hythe Formation is exposed in all the lane sections listed above. The most interesting section is at Nutbourne [TQ 07311822], where the Hythe Formation (like that seen at Warmingshurst) may be observed. In addition, the most southerly part of the road cutting exposes a grey waxy clay in place of a hassock bed. This clay is rich in glauconite and a clay mineral of the smectite group, calcium montmorillonite. It is typical of the fuller's earths described by Young & Morgan (1981) in the Hythe Formation of the South Downs. 5.5 m to 6 m above the lowest Hythe Formation exposed at Nutbourne a pebbly clay, with silica-cemented white sands, marks the base of the overlying Sandgate Formation. At the top of the Nutbourne road cutting described above, a typical (calcareous matrix) 'Bargate Beds' (now termed Bargate Member - see below) fauna may be collected from what is presumably the lowest part of the Fittleworth 'Beds'. This includes *Resatrix parva*, *Nanonavis*, *Linotrigonia*, and *Mesolinga*, echinoid spines and the brachiopod *Sellithyris*.

Stop 4. Folkestone Formation, Washington Sandpit [TQ125135 to 124137].

How to get there
    The pits to the south of Rock Crossroads, Washington have long been a source of quality building sand. These workings have removed an overburden of Gault Clay in order to extract the sand, providing a section

Early Cretaceous Environments of the Weald

through the Folkestone - Gault Formation boundary. The boundary may be viewed without recourse to entering the pit by joining the public footpath just east of Washington [TQ 124131] and walking through the abandoned pits to the A24. These may become overgrown when quarrying is not taking place.

Key literature
Anderson (1986).

Summary of geological interest
Folkestone - Gault junction and the Iron Grit.

Description
The key interest at Washington is the Folkestone Formation - Gault (Clay) contact. Here, the boundary beds are developed in the "Iron Grit" facies, contrasting with the phosphate nodules found at the same horizon at Folkestone. The Iron Grit defines the dip of the beds here as being 5° - 10° to the south. This is important as true bedding in the cross-stratified Folkestone Formation below may be difficult to discern, whilst bedding is not obvious in the slumped and homogenous Gault Clay.

The Iron Grit itself varies from 20 cm to 1 m in thickness and comprises rounded quartz pebbles and limonite ooliths with rare rock clasts in an iron oxide cement. Below, the Folkestone Formation consists of about 30 m of 1 m to 2 m thick cross-stratified sets of medium-grain size sand, although it is interesting to note that the sets become thinner (40 cm to 1 m), heavily iron-stained and intensely burrowed by the trace fossil *Ophiomorpha* (a crustacean) in the 2 m of section below the Iron Grit (Figure 23). Cross-stratified sets tend to have sharp upper and lower contacts with shallow water vertical *Skolithos* burrows penetrating down from individual foresets.

"Environmental interpretation of the Iron Grit is difficult" (Wach & Ruffell, 1991). Anderson (1986) [considered] a lagoonal origin. It would be best to remember that this condensed bed is most likely equivalent to the Carstone Formation of the Isle of Wight and could represent not only a long time gap, but several depositional - erosive episodes. The existence of variably cross-stratified sands confirm periodic changes in current energy in the Folkestone Formation below. The sands were deposited within a shallow sea, swept by strong tidal currents with a dominant flow towards the south and southeast.

Early Cretaceous Environments of the Weald

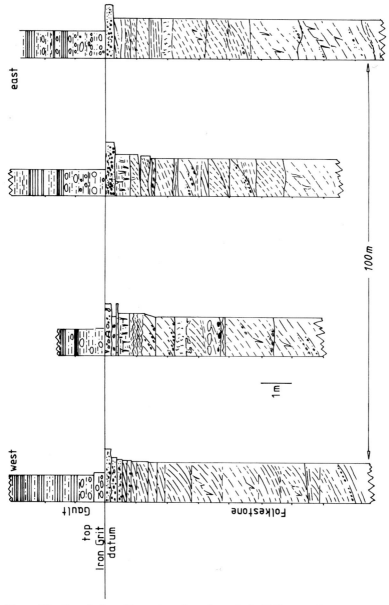

Figure 23. Correlation of logged sections through the Folkestone Formation - Iron Grit - Gault Clay at Washington Sandpit. Key as in Figure 26.

Early Cretaceous Environments of the Weald

Stops 5. Sandgate Formation from Marehill [TQ 06451878] to 6. Pulborough [TQ 04751870] and 7. Fittleworth [TQ 01431954]. (Sheet 197).

How to get there
All the locations exposing sediments of the typical Sandgate Formation of the South Downs can be examined along the line of the A283. These outcrops are effectively either side of the railway station at Pulborough, from where a bus service operates to the east or west. The A283 is a busy road and no public footpath traverses the outcrops easily.

Key literature
Humphries (1964) gives stratigraphic sections of the Lower Greensand: the fossiliferous location of Park Lane, Pulborough is mentioned in Bristow & Morter (1983).

Summary of Sandgate Formation stratigraphy
South Downs Fittleworth - Rogate 'Beds'
As Humphries (1964) and Bristow and Wyatt (1983) noted, there are few exposures of the lowest Sandgate Beds (now Formation). The Fittleworth Beds (now Member) can be examined at the rather degraded type locality [TQ 01431954] and consist of five metres of grey and green bioturbated silty sands. Estimates of the Fittleworth thickness are 30 to 52 metres around the type locality. Further east, the divisions of the Sandgate Formation cannot be recognised and the whole of the Sandgate Formation has thinned to 18 metres in the Washington Borehole and to 8 metres in the Nep Town and Streat Boreholes. The Lower Greensand Group is so thin as to be undifferentiated in the Hampden Park Borehole in Eastbourne. Unfortunately, core recovery in the I.G.S. and B.G.S. boreholes along the South Downs is more intermittent in the Sandgate Formation compared to the Hythe Formation, on account of the differences in cementation. Around Selham, Wooldridge (1947) and later Humphries (1964) mapped a separate unit within the Rogate Beds (now Member), the Selham Ironshot Sands. The limonite pebbles that constitute this unit are confined to a relatively discrete layer, such pebbles occurring throughout the Sandgate Formation of the Western Weald as isolated clasts and minor pebble beds. The Selham Ironshot Sands in the type locality are about 5 metres thick and the limonitic Rogate Member, around Petersfield, is about 40 metres thick. North of here the topmost Rogate Member become clay rich, a trend that continues towards the North Downs.

Pulborough Sandrock
Above the poorly exposed Rogate and Fittleworth Formations the Sandgate Formation comprises the Pulborough Sandrock (*cunningtoni* subzone,

Early Cretaceous Environments of the Weald

nutfieldiensis zone) and the Marehill Clay (*nolani* subzone, *jacobi* zone) overlain abruptly by the Folkestone Formation. Both are considered members as they possess very distinctive lithologies, easily recognised in the field and in boreholes. The Pulborough Sandrock is represented by a layer of nodules in the east (Washington Borehole) and 35 metres of sandstone in the west. At outcrop it consists of a yellow (brown weathering), soft, occasionally iron-cemented sandstone. Ironstone bands and nodules preserve a characteristic fauna: the fossils to be found are outlined at each location. The lack of outcrop and borehole data makes correlation of the Pulborough Sandrock northwards (into the Puttenham Member) somewhat hazardous; Middlemiss (1961a) found a Pulborough Sandrock lithology within the Puttenham Member at Headley [TQ 814372]. The Pulborough Sandrock is the same age as Group XIV on the Isle of Wight (Ferruginous Bands of Blackgang Chine). Much of the data used here is from Bristow & Wyatt (1983) and Bristow (1981; unpubl.), for the Pulborough/ Storrington and Petersfield areas respectively.

Marehill Clay
This dark silt- and clay-dominated unit abruptly overlies the Pulborough Sandrock at all localities, occasionally with a pelletal mudstone-siltstone bed, but more usually with an iron-pan developed at the top of the Pulborough Sandrock, and indications of very strong leaching of the sands below. The overall tendency of the Pulborough Sandrock to coarsen upwards suggests a significant hiatus before deposition of the Marehill Clay. The dark colour of the Marehill Clay is due to organic content; the occasional dark green colour is ascribed to glauconite. In thick successions such glauconitic middle parts of the member pass into a Pulborough Sandrock-like facies, but this has only been recorded in boreholes and is difficult to interpret sedimentologically. The Pulborough Sandrock may be a marginal facies of the Marehill Clay. Parts of the Marehill Clay are well laminated with large mica flakes on bedding planes. Conversely, in the more homogenous successions, no biogenic structures can be detected, the occasional white silty and fine sand laminae being completely undisturbed. Surprisingly, no fauna has ever been recorded from this member and washed residues are devoid of calcareous microfossils. The palynofacies has been examined by D.J.Batten who found the residue to be dominated by coarse plant fragments, vitrinite, common miospores and dinocysts, indicative of a late Aptian nearshore, marine environment. Although the Marehill Clay varies in thickness along the South Downs the clay content stays very much the same, though vertical sampling shows some kaolinite increase upwards. Wood (1957) studied the heavy minerals and found them to be very similar to the Pulborough Sandrock, and without much lateral variation. This data suggests a complete "blanketing" of the Portsdown area in early *jacobi* times by a silty mudstone with some organic content, but no obvious fauna. The stratigraphy of the Sandgate Formation in this area is summarised in Figure 24.

Early Cretaceous Environments of the Weald

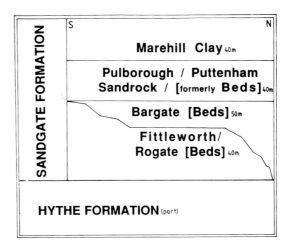

Figure 24. Stratigraphy of the Sandgate Formation in the South Downs.

Stop 8. Church Road, Pulborough [TQ 04751875].

Description

In the road cutting the junction of the Pulborough Sandrock and Marehill Clay can be seen dipping at around 10° to the southwest.
The Pulborough Sandrock is characterised here by the presence of glauconite, the trace fossil *Ophiomorpha*, some tabular cross-stratification, but more commonly current-ripple lamination. It is only as drapes on such sedimentary structures that clay is preserved in this section, although some rip-up clasts (of former clay drapes?) are found in the middle of the Pulborough Sandrock both here and in the Fittleworth outcrops (see below). The Marehill Clay forms most of the bank along the road, and is developed in typical dark, silty and glauconitic facies.

Stop 9. Park Lane, Pulborough [TQ 03951890].

Description

Exposures of the Pulborough Sandrock occur in the banks of Park Lane; these never show more than a metre of vertical section, yet are highly fossiliferous. The bivalve fauna shown in Figure 25 also includes *Gervillella*, *Nucula*, *Parmicorbula*, *Resatrix*, *Arca* and the rudist *Toucasia*. The last form is a Tethyan immigrant, indicative of warm waters, and unusual in English Cretaceous sediments. The fauna is exclusively preserved as moulds showing internal features in exquisite detail. Never, however, can the collector find a removable body fossil or mould. Very similar exposures occur along Park Lane

Early Cretaceous Environments of the Weald

at TQ's 03951893; 03961897; a Pulborough Sandrock fauna may also be found in trackside exposures in Northpark Wood, Parham Park [TQ 05351525].

A series of minor stops: Hesworth Common [TQ 00601933, 00621934 00401942] and Fittleworth [TQ 00821927].

Description
The Pulborough Sandrock at these locations comprises medium and fine grained sands with a high clay content, sometimes developed as grey and lilac coloured drapes on ripples and bedding planes. Bioturbation is very common to the point of being completely pervasive and occasional branching burrows of *Thalassinoides* type can be distinguished. Ironstone nodules have preserved the moulds of a shelly fauna. These are collectable at all the outcrops around Fittleworth, and include *Thetironia minor*, *Corbula striatula*, *Nucula* and *Anchura* (*Perissoptera*) *robaldina*. Sieved residues of clay drapes yield siliceous (elongate) sponge spicules and ostracods. Such nodules and "pans" also preserve a rare, remarkable fauna (listed by Casey 1961, and Bristow & Morter 1983), the most notable part of which is the rudist *Toucasia*.

Where there is an absence of clay, "floating" limonite ooids in a fine sand matrix can be observed. Where bedding is visible, the concentration of ooids varies in graded (fining-up) successions of around 50 cm thickness. Although strongly bioturbated, a few distinct biogenic structures can be seen and in the most easterly exposure of Hesworth Common [TQ 00401942] the thin worm burrow *Planolites* can be identified. The limonite ooids may have been derived from the Portsdown Swell, either forming penecontemporaneously along with the limonitisation of glauconite or as reworked material from the limonitic pebbles of the underlying Sandgate Formation. As the two sources could be different in age by less than an ammonite subzone, the fresh state of the ooids is no guarantee as to the age of its formation. Very rarely in the uppermost and lowermost beds of the Pulborough Sandrock one can find complex grains of limonite ooids with a phosphate overgrowth, or alternating glauconite and limonite rings within ooliths. Often such grains have ?fungal borings and preserve foraminifera and, in contrast to the simple limonite ooids, could have had a much more complex history of reworking in a shoal environment (or on adjacent "highs"). The iron cementation that preserves the shelly fauna described above is commonly found in association with such iron oolite "shoals".

Stop 10. Bognor Common Quarry [TQ 00802135].
    (Sheet 197)

How to get there
    Bognor Common is traversed by many footpaths and bridleways, only

Early Cretaceous Environments of the Weald

accessible by private transport or a long walk from the infrequent A283 buses. The quarry is remarkable in being situated within the common, with no restrictions on access. Nonetheless, groups or those carrying out detailed fieldwork should contact the owners (The Local Stone Company, Bognor Common) to check on access.

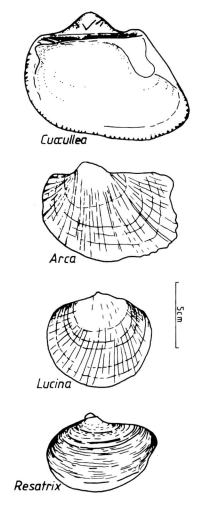

Figure 25. Common bivalves found in the Hythe Formation. Most are preserved only as casts on siltstone - sandstone bedding planes.

Early Cretaceous Environments of the Weald

Key literature
Young & Morgan (1981); Bristow & Morter (1983); Ruffell (1992).

Summary of geological interest
Bivalve fossils; silica and carbonate diagenesis; fuller's earth; shallow marine sedimentary structures.

Description
The published accounts give very adequate descriptions of the Hythe Formation at Bognor Common. Bedding is very obvious by virtue of silica cemented bands, around 10 cm to 50 cm thick, interstratified with laminated and uncemented muds, silts and fine sands. Ruffell considered the silica cement to occur preferentially in clean, white sands. Sedimentary structures may be difficult to discern because of this cement. The clearest example of trough cross-stratification can be seen in the lowest beds of the quarry (close to the actively worked floor), where uncemented sands, silts and waxy clays infill broad (10 m wide) channels.

Ruffell also described the bivalve fossils in two lithological associations at Bognor Common. The first, representative of the *bowerbanki* zone of the Lower Aptian, was listed by Bristow & Morter (1983), and includes *Aetostreon*, *Trigonia*, *Resatrix* and *Mesolinga*, these being evident on sandstone - siltstone/ mudstone bedding interfaces. The second is found exclusively in the montmorillonite-rich mudstones (fuller's earths) described by Young & Morgan (1981), and includes *Cuccullea*, *Arca*, *Inoceramus* and *Lucina*. (Figure 24).

Stop 11. Disused Folkestone Formation Sandpit, Champs Hill Camp-site, Codwaltham [TQ 02151640].
(Sheet 197).

How to get there
At the southwestern end of Codwaltham village the A29 Pulborough - Chichester Road intersects a cross-roads. The minor road that passes northwest (the Fittleworth Road) has an iron gate adjacent to some stables ("The Pines") some 50 metres from the cross-roads. The iron gate leads to Champs Hill disused sandpit (now a village camp-site and recreation ground). Although this area is parkland, groups should seek permission to pass through the gateway at "The Pines".

Summary of geological interest
Tabular cross-stratification, vertical trace fossils, clay drapes.

Description
Although there are no published details for this section, the Folkestone Formation sands are very typical of their equivalents throughout the western

Early Cretaceous Environments of the Weald

Weald. In this location weathering and careful curation by the local community has preserved one of the best permanent exposures of large-scale tabular cross-stratification available south of London. The 12 to 14 metre high face shows five obvious sets of cross-strata. Both the horizontal and cross-stratified beds are picked out by clay beds that commonly "drape" rippled and inclined beds. Such clay-drapes are common indicators of tidal activity in the marine environment in which these beds formed. Each cross-set was, in effect a migrating sandwave, sometimes 2 to 3 metres in height and looking much like a submarine dune. Tidal currents pushed this sandwave through the Cretaceous seaway that filled the Weald Basin. From the inclination of the cross-stratification (towards the south and east) we may conjecture that the dominant tidal currents also moved in those directions.

N.B. Please do not hammer or scrape the faces here. The location is both a public amenity, scout camp and memorial to a local notability, Alfred Bowerman. Similarly, the cave (in trees) at the western end of the main pit should not be entered.

Itinerary 6. The core of the Weald Anticline around Petersfield
Alastair Ruffell

Unlike the other itineraries, there are no major outcrop locations around Petersfield area. Instead, our knowledge of the Cretaceous succession has been built up through the study and mapping of numerous small exposures. This part of the Guide differs from the rest in being written as a travelog, beginning 6 km east of Petersfield, near Rogate and ending 4 km northeast at Hill Brow. All the localities are on **Sheet 197, 1:50,000.**

Habin River Bridge, near Rogate [SU 808229]. Take the minor road running S from the A272 at Rogate. Exposures of the Pulborough Sandrock (towards the top of the Sandgate Formation), a well lithified, pale yellow, fine grained sandstone, carrying relatively shallow water marine fossils occur here.

River Rother footbridge, Habin [SU 80282286]. Walk along the riverside footpath to this location, which is about 400 m west of Habin River Bridge, for further exposures of the Pulborough Sandrock. The faces exposed are rarely greater than 2 m in height.

A272 roadside exposures [SU 80252360]. This is another small exposure of the Pulborough Sandrock, located about 400 m west of the cross-roads in Rogate.

West Heath Common Sandpit [SU 78502265]. Access to this sandpit and to the railway cutting (see below) is via a public footpath running eastwards, just

Early Cretaceous Environments of the Weald

200 m north of Rival Lodge. West Heath Common is 2 km WSW of Rogate and is traversed by trackways and bridlepaths cut into shallow exposures of the marine Folkestone (sands) Formation. The best long-lasting exposure is in the abandoned sandpit (centred on SU 784228). Loose fragments of ironstone ("Carstone") are scattered everywhere.

Disused railway cutting [SU 78442318 to SU 78302350]. In the old railway cutting fringing the northern side of West Heath Common, Pulborough Sandrock blocks can be found. Be warned that the cutting fills with water after heavy rain.

A272 road cutting, Durleighmarsh [SU 77802375]. This somewhat overgrown cutting exposes the Selham Ironshot Sands, a medium grained, cross-bedded deposit carrying black grains of limonite. It occurs at a slightly lower level within the Sandgate Formation succession than the Pulborough Sandrock, but it does demonstrate the variability in the lithology of the Sandgate Formation as a whole in this area (Humphries, 1964). Throughout the Weald, changes in lithology and thickness typify the Sandgate Formation, suggestive of nearshore irregular patterns of deposition controlled by the local sea-bottom and adjacent coastal physiography.

River Rother Bridge, Sheet (Petersfield) [SU 76242450]. This bridge carries the A272 and nearby occur exposures of calcareous and silica-cemented Hythe Formation beds. The silica cementation was probably brought about by the dissolution of marine siliceous organisms, especially sponges, within the soft bottom sediment followed by precipitation of the dissolved silica from the pore fluids.

Steep Woods - Bedales School [SU 74782518]. Abandoned sandpits are now overgrown with scrub oak and brambles, yet there are many shallow exposures in the Folkestone (sands) Formation.

Old Sand Pits, Princes Marsh, near Liss [SU 76072519 - SU 75632537]. Fossiliferous Pulborough Sandrock with numerous moulds of the marine bivalves *Corbula, Cuccullea, Thetis* and *Gervillella*.

Stodham Park [SU 772262]. Take the minor road leading northwestwards from the B2070, about 1.5 km from the A272 junction. In the laneside banks of the village, occur pebbly, uncemented lower beds of the Sandgate Formation (Rogate Member). These rest directly on uncemented sands of the topmost Hythe Formation, exposed at SU 77652638.

Brow Hill [SU 80002595]. The topmost Hythe Formation here comprises slumped cross-stratified beds described by Humphries (1964), Middlemiss (1961b, described at Combe Hill, SU 801258) and Ruffell (1992). This locality

Early Cretaceous Environments of the Weald

is very good for showing the instablility of sediment during sandwave formation. The sheer forces engengered by powerful bottom currents appear to have been such as to lead to the mobilisation and mass movement (slumping) of the partially compacted sands.

Itinerary 7. Haslemere to Godalming
Alastair Ruffell

(Sheet 186, 1:50,000)

How to get there

The three main stops and one minor stop of this excursion are all widely separated. To visit all three together private transport is recommended. The Devil's Punch Bowl may be visited by bus as it is adjacent to the A3; Stock Farm Quarry is very isolated with no public transport; Brook is on a bus route serving the Institute of Oceanographic Sciences at Wormley; Milford Cemetery may be approached from the A3 bus routes or from Milford railway station.

Key literature

Humphries (1964); Kirkaldy (1933); Kirkaldy & Wooldridge (1938); Kirkaldy & Middlemiss (1954); Knowles & Middlemiss (1958); Middlemiss (1962b); Richardson (1947); Thurrell, Worssam & Edmonds (1968).

Summary of geological interest

From the core of the Weald anticline around Petersfield to the North Downs the various formations of the Lower Greensand Group show significant facies changes. Most of these changes are greatest in the complex outcrops around Haslemere. The volume of literature reflects the effort put into understanding this area by both the Weald Research Committee of the Geologists' Association and the Geological Survey in the inter- and post-war periods. The specialist is recommended to read this literature. Many of the exposures are small, isolated, but luckily seem to have survived the passage of time, many by virtue of being within nature reserves or National Trust property. We shall concentrate here on three main locations.

Stop 1. Devil's Punch Bowl [SU 90003580].

How to get there

Although the area is a nature trail, it is traversed by the very busy A3, along which buses from London to Portsmouth pass. A car park at SU 89283577 is the most convenient starting point, from where the visitor

Early Cretaceous Environments of the Weald

can take a path east (parallel to the A3) to the viewpoint or go north to Highcomb Bottom, where the main geological interest lies.

Key literature
Knowles & Middlemiss (1958); Middlemiss (1961b); Thurrell, Worssam & Edmonds (1968).

Summary of geological interest
Around the Devil's Punchbowl deep erosion along the numerous valleys of the area has incised through the capping of uppermost Hythe and Sandgate Formations to expose the Hythe Formation. Unfortunately the outcrops are isolated, prone to slippage, and covered with vegetation, otherwise the area would be a Hythe Formation type section.

Description
The numerous Hythe and Sandgate Formation exposures in the valleys draining Gibbet Hill [SU 8998 3582] cannot be covered comprehensively here. Instead, the features found in the best section, at Highcomb Bottom, are described and it is recommended that the interested reader use this as a demonstration section, whence to explore the other outcrops.

In the river valley adjacent to a small patch of grazing land in the lowest area of Highcomb Bottom [SU 89303676] the fine grained sand with clay matrix characterising the lowest part of the Hythe Formation can be observed (Knowles & Middlemiss, 1958, p. 210). As one ascends Highcomb Bottom, toward Gibbet Hill, numerous Hythe Formation exposures show the beds as both coarsening-up and increasing in glauconite content. At the top of the gorge, adjacent to the A3 main road, interbedded glauconite and quartzose sands can be seen. These pebbly beds are most likely at the top of the Hythe Formation in this area.

The Cenozoic and recent incision of river valleys in the area is of some interest. This process is uncommon in the Weald, and it is interesting that the Haslemere - Petersfield area is at the core of the Weald Anticline, a feature created by Cretaceous - Cenozoic basin inversion, with the final uplift in Miocene times. The fluvial incision probably dates back at least to the Miocene, and as incision is continuing today then it is reasonable to say that post-Alpine uplift may still be ongoing in the area.

Stop 2. Stock Farm Quarry, Churt [SU 87673829].

How to get there
Stock Farm Quarry is disused, and being located on the small road

Early Cretaceous Environments of the Weald

between Beacon Hill (SU 87353666) and Rushmoor (SU 87454030) away from major conurbations has not been used extensively for landfill. The quarry entrance (a gateway) can be easily seen from the road.

Key literature
Richardson (1947); Knowles & Middlemiss (1958); Thurrell *et al.*(1968).

Summary geological interest
Bargate (Beds) Member; reworked and *in situ* fossils; large-scale cross-stratification.

Geology (background)
Lack of biostratigraphical control has resulted in much confusion over the Sandgate (Beds) Formation succession in this area, especially in the centre of the Weald Anticline where the large thickness of the formation and the somewhat gradational passage between individual members makes mapping difficult. The pebbly and cross-stratified lower parts of the Sandgate (Fittleworth/ Rogate Beds correlative) have an intermittently developed calcareous cement throughout this area and towards the North Downs, where they are referred to as the Bargate beds (now a member of the Sandgate Formation). Large concretionary masses ("doggers") appearing sporadically above the cross-bedded Upper Hythe beds, especially in the Hindhead area, form a gradational boundary between Hythe and Sandgate Formations. Passing onto the basin margins a clearer break between these two is visible, mostly through pre-*nutfieldiensis* erosion of the Upper Hythe beds. In this area the Bargate beds are characterised by the abundance of derived Upper Jurassic fossils they contain (Arkell 1939). As one passes east along the North Downs one can find the Bargate beds pebbly units resting on the Top or Mid-Hythe Pebble Beds, making for much confusion.

Description
Stock Farm Quarry is the best exposure in the area and, although disused, it still shows 20 metres of Bargate beds. Richardson noted cherts in the upper, cream-coloured layers, whilst today the base of the overlying Puttenham beds can be seen at the very top (Figure 26). The beds at the base of the quarry appear to be lithologically similar to those above, but differ subtly with the inclusion of rare Upper Jurassic ammonite steinkerns (internal moulds), some pebbly layers and much comminuted shell debris. The 8 to 9 metres of section above comprise alternating chertified sandstones and glauconitic clayey siltstones, coarser at the base and top, with massive tabular to shallow trough cross-stratification (Figure 26). Palaeocurrent vectors suggest movement of these small sandwaves toward the south most of the time. The Puttenham beds at the top of the quarry are transitional with the underlying Bargate beds and contain the trace fossil

Early Cretaceous Environments of the Weald

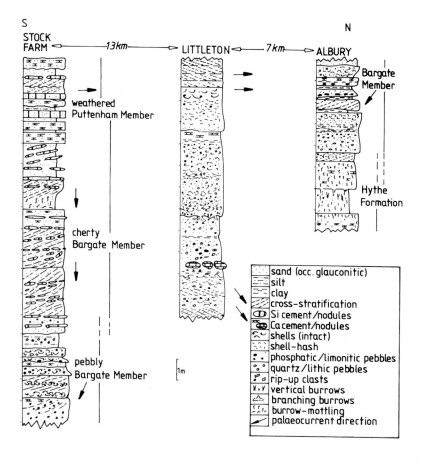

Figure 26. Correlation of the pebbly Sandgate Formation ('Bargate Beds') of the western and northern Weald.

Early Cretaceous Environments of the Weald

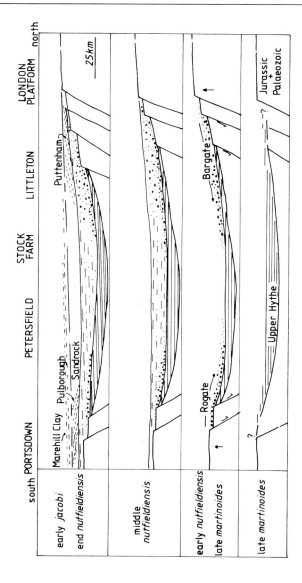

Figure 27. Cartoon of the tectonic movements thought to have controlled deposition from late Hythe Formation deposition ('martinioides zone') to late Sandgate Formation deposition ('early jacobi zone') across the western margin of the Weald Basin.

Early Cretaceous Environments of the Weald

Macaronichnus. From an examination of the isopach maps published by Sellwood, Scott & Lunn (1986) and Ruffell (1992), the Stock Farm Quarry area is situated in the very depocentre and presumably most rapidly subsiding part of the Weald Basin in Cretaceous times. The rarity of derived clasts was commented upon by Knowles & Middlemiss (1958), who rightly suggested that the area was distant from the London Platform source-lands; northwards the volume of reworked Jurassic material progressively increases (Figure 27). The effects of this influx of material can be seen in the next good Bargate beds exposures at Littleton [around SU982474], which were examined by Kirkaldy (1933). Here, although still calcareous and shelly, the beds form part of a thinner succession with much more clearly defined pebble beds, large glauconite grains and a varied assemblage of fossils, from terebratulid and rhynchonellid brachiopods to the ubiquitous bivalves *Aetostreon*, *Modiolus*, *Pectunculus*, *Limopsis*, and small sponges, bryozoa and regular echinoid spines. All these organisms were indigenous to the Cretaceous sea-way of the area.

Stop 3. Brook Road Section [SU 93003800 to 93093850].

How to get there
    This stop is intended as a short "look and see" locality as the road (A286) can be busy, and there are no convenient access points. The lone walker can pass along the verge for the length of the section. Those with private transport may either park in Brook (the "Dog & Trumpet" is a recommended geological watering hole), or drive up the road section *en route* for Milford, or down to Stock Farm/Devil's Punchbowl. Groups should not stop at this section.

Key literature
    Knowles & Middlemiss (1958); Ruffell (1992).

Summary of geological interest
    Condensed Hythe Formation facies; chert cementation.

Description
The southernmost, and stratigraphically lowest Hythe Formation beds to be exposed in the road cutting north of Brook are uncemented, fine-grained and bioturbated sands. These are the "Fine Sands of Witley" described by Knowles & Middlemiss (1958): they contrast markedly with the beds above, these being dominated by a chert cement (the "Cherty Division" of Knowles & Middlemiss (1958)).

Stop 4. Milford Cemetery, Witley Common. [SU 93854165].

How to get there
    Witley Common is National Trust property, for which there are car-parks

Early Cretaceous Environments of the Weald

on both the west side of the A286 [SU 94204147] and on the east side (marked on O.S. Sheet 186 at SU 83674080). Access is open, although no activity which destroys the trust's property (includes digging) is allowed. If in doubt concerning serious geological investigation, the trust should be consulted. From either car-park, the visitor should walk 200 m northeast to the boundary wall of Milford Cemetery. Within the wooded area to the south of the wall is a 40 m - 50 m diameter sand-pit, in the upper slopes of which the Puttenham Member is exposed.

Key literature
Middlemiss (1962b).

Summary of geological interest
Massive tabular cross-stratification; glauconitic sands; burrows of the "Vermiform" (*sensu* Middlemiss) or *Macronichnus* type.

Description
Two aspects of Lower Greensand geology are striking at Milford Cemetery, the relatively large scale of the (Puttenham [Beds] Member) cross-strata and the clarity of preservation of the *Macronichnus* burrows. The sands at Milford are of fine to medium grain-size, with variable, yet ubiquitous glauconite content (10% - 40%). This results in the sands appearing very dark, and when bioturbated having a 'salt and pepper' appearance. The nature of the cross-stratification indicates that you are looking at cross-sections through migratory sandwaves. The inclined cross-strata (foresets) dip to the south and east, implying that this was the likely direction of sandwave movement.

The burrows, here termed *Macronichnus*, were described by Middlemiss (1962b) as "vermiform" as after *Vermes* (worms) and his description requires no addition here.

Itinerary 8. The Hog's Back (Farnham to Dorking)
Alastair Ruffell

(Sheets 186 & 187, 1:50,000)

How to get there
Farnham, Guildford and Dorking all have excellent train connections for London, although only a few of the stops are easily accessible direct from these towns. The A31 Hog's Back road is a major route for buses from Guildford to Farnham, from which all the remaining stops may be reached on foot.

Early Cretaceous Environments of the Weald

Key literature
Arkell (1939); Dines & Edmunds (1929); Middlemiss (1978); Lake & Shephard-Thorn (1985).

Summary geological interest
Derived fossils; tidal sedimentology; inversion tectonics.

Description
The structural geology of the North Downs is dominated by the Hog's Back (and associated) folds. Here, we see not only the outcrop expression of structure in the rocks deformed by Alpine (Cenozoic) processes, but also evidence of the earlier uplifts to have affected the area: the inclusion in the Lower Greensand strata of worn, rolled and transported (derived) Jurassic fossils was brought about by uplift of the London Platform to the north in Aptian times.

The visitor who wishes to see more in this area than just the representative outcrops included here, is advised to use Lake & Shephard-Thorn (1985), whose comprehensive account provides details of all major outcrops.

Stop 1. Mear's Pit, Runfold, Farnham [Around SU 865476].
(Sheet 186)

How to get there
On the A31 (Hog's Back Road), 2 km east of Farnham town centre, there is an area of barren ground before the hamlets of Runfold and Sandy Cross. In this area extraction of quality sand has occurred for many years (Shepherd, 1934). All the pits are actively used as inert waste dumping grounds, and the state (and fate) of exposures is highly variable. From our own experience, and from the word of the operators, only the upper beds in Mear's Pit are likely to survive for the foreseeable future. The entrance is on the north side of the A31, where there is a wide gateway (with parked cars on weekdays). The beds of interest are adjacent to this gateway, and in the immediately adjacent entrance to the quarry. Access to the quarry itself is restricted, the floor being flooded.

Key literature
Casey (1961); Narayan (1971); Padgham (1972) & Shepherd (1934).

Summary geological interest
Gault - Folkestone Formation junction; intra-Folkestone Formation clay beds; sandwave sedimentology.

Description
Two major features are of interest in Mears Pit: the Gault - Folkestone

Early Cretaceous Environments of the Weald

Formation junction and the Folkestone Formation. In the small cutting that forms the entrance to the quarry the Gault Clay is visible, resting on Folkestone sands. Only the lower 3m of the Gault is exposed here, and consist of black silty sands with a clay matrix. The beds are thoroughly bioturbated and bedding is only visible as a line of 5 cm - 10 cm diameter white, gritty nodules, about 50 cm above the top of the Folkestone sands. The Gault is fossiliferous, ammonites and gastropods occurring both here and in any dug material between the pit and the disused railway to the north. Fresh surfaces may show polished and slickensided fabrics and this may be a product of the faulting described by Lake & Shephard-Thorn (1985). Neither the Carstone nor the black phosphate nodules characteristic of this junction at other locations throughout the Weald, are visible here.

The Folkestone sands around Farnham are yellow/orange (with iron cement) or white/silver (with no iron cement). Confusion has arisen from the iron-cemented, gritty layers within the Folkestone sands having been traditionally termed "carstone". This term has a stratigraphic significance as a sedimentary unit (not a lithology) - the Carstone Formation of the Isle of Wight and Norfolk. Iron-cemented, gritty layers commonly form at the boundaries between cross-stratified units (Figure 28) and are informally termed 'iron-pan'. One such layer is clay-rich and contains the calcareous, fossiliferous nodules described by Shepherd (1934) and Casey (1961). This layer, from which the *farnhamensis* fauna was described, is now generally covered at the bottom of Mear's Pit.

Stop 2. Thorncombe Street, Godalming [SU99784242].
(Sheet 186)

How to get there
Thorncombe Street is a road cutting in the Hythe Formation to the southeast of Godalming. The road itself is minor, adjacent to the B2130. The section is not easily accessible without private transport, being just over 3 km from Godalming Station.

Key literature
Kirkaldy (1932); Ruffell (1992).

Summary geological interest
Silica and carbonate cemented Hythe Formation; comparison of Hythe - Sandgate contact with successive locations (Albury).

Description
Of the four major Hythe Formation outcrops around Godalming, Thorncombe Street is the most accessible and pleasant to work at. The other three are at: Busbridge, SU 98004226 (mentioned in Ruffell, 1992), Winkworth Farm,

Early Cretaceous Environments of the Weald

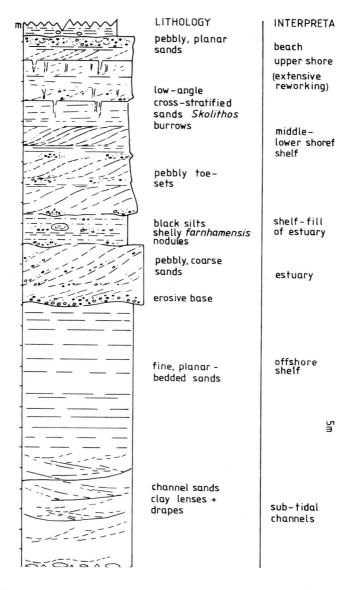

Figure 28. Stratigraphy of the Folkestone Formation around Farnham, including Mear's Pit, Runfold. Horizontal width expanded to show features of cross-stratification.

Early Cretaceous Environments of the Weald

around SU 995407 (mentioned in Kirkaldy, 1932), and the new Hythe Formation quarry at Unstead Park (likely to be short-lived), around SU 988443. A log of the beds at Thorncombe Street is given in Ruffell (1992), from where it can be seen that the beds show cemented - uncemented alternations similar to the 'Rag & Hassock' of Kent and Sussex. At the north western end of the road cutting at Thorncombe Street, very fossiliferous Bargate "Beds" (now a member within the lowest Sandgate Formation) may be found. The contact with the Hythe Formation is not visible unless the lane is cleared (as it is on occasion by farmers).

Stop 3. Waterloo Pond, Albury [TQ 04134807].
(Sheet 186)

How to get there
This outcrop of the lower parts of the Hythe Formation is on the north side of Waterloo Ponds, about 250 m north of the A248 Shalford - Dorking Road.

Key literature
Lake & Shephard-Thorn (1985).

Summary geological interest
One of many small Hythe Formation exposures in the vicinity, included for proximity to Albury and the main road, and introduced because the lowest Hythe Formation is rarely exposed in this area.

Description
The Hythe Formation here contrasts markedly with the Godalming exposures, being poorly cemented, fine to medium sand with around 10% glauconite.

Itinerary. 9 Albury
Alastair Ruffell

(Sheet 186 1:50,000)

How to get there
Albury is situated on a marked double-bend in the A248 about 5 km ESE of Guildford. The road from Chilworth Station is very narrow and dangerous for pedestrians.

Key Literature
Kirkaldy (1976), Lake & Shepherd-Thorn (1985).

Early Cretaceous Environments of the Weald

Summary of geological interest
Hythe Formation beds and Bargate Member with derived phosphatised Jurassic fossils. Cross-bedded Folkestone Formation sands. Classic viewpoint across the Weald.

Stop 1. Great Halfpenny Lane [TQ 02154842 to TQ02104817].
A good section in a road-cutting, on a minor road 2 km N of Chilworth, through the pebble beds at the base of the Folkestone Formation. Sometimes the contact with the underlying Bargate beds is exposed.

Stop 2. Guildford Lane [TQ 0465 4805].
Sections through the Hythe and Bargate beds in a road-cutting just off the A248 (see below).

Stop 3. Blackheath Lane, Albury [TQ 04804743 to 04824743].
This is a very minor, partly 'sunken lane', leading due S from the A248 at Albury. The Hythe Formation occurs towards the southern end of the traverse. Here, it is a fine to medium grained sand carrying glauconite, indicating its true marine origin. Some of the beds, as much as 40 cm thick, are strongly calcite-cemented imparting a kind of cyclicity to the beds, such as one also sees in the 'Rag and Hassock' facies further eastwards in Kent (see p.73 and Middlemiss, 1975). Rare quartzite pebbles and occasional brown-black phosphatic pebbles can be found 'floating' in the sands, some of which are bioturbated. A higher degree of calcite cementation and increase in the number of pebbles as one moves northwards along the lane, indicate the transition into the Bargate Member beds. Precisely the same transition can be observed in Guildford Lane. The phosphatised pebbles, and some limonitised moulds (steinkerns), are of great interest in comprising Jurassic Fossils. They provide evidence that during late Hythe and Bargate times there must have been some uplift and erosion of Jurassic strata on the London Platform to the north (Figure 27) (Arkell, 1939).

(Note. For futher exposures of the Hythe Formation and Bargate Member the ground east of Albury has numerous north-south 'sunken lanes' leading from the A25, some transversing Abinger Common and Wotton Common. These lanes cut through the beds and there are many outcrops, some being listed in Lake and Shephard-Thorn (1985)).

Stop 4. Albury Lane Sandpits [TQ 04724850 to TQ 05514842].

How to get there
From the A248 there is a dead-end lane (Albury Lane) leading northwards towards the Downs. About 400 m up the lane are sand pits in the Folkestone Formation. Permission to enter is difficult to obtain due to

Early Cretaceous Environments of the Weald

periodic extraction, but a verbal communication with the operator, if on site, is usually all that is ever needed. Access to the pits, can also be from the A25 at TQ 05574865, taking a footpath leading west.

Key literature
Lake & Shephard-Thorn (1985), Middlemiss (1975).

Summary geological interest
The variable lithology of the Folkestone Formation.

Description
Lake and Shephard-Thorn summarise the Folkestone sands at Albury as "30 m of cross-bedded sands with carstone, dipping at 20 degrees N". The relatively high dip probably relates to the proximity of the Hog's Back Monocline, the eastern continuation of which passes via Guildford through to Dorking. The full thickness is not exposed in the pits, but they do show cross-sections representative of the variable lithology, with silver, red and yellow sands spectacularly cross-bedded and rippled in places. Studies of the direction of inclination of the forests, here and elsewhere along the north crop of the Weald, indicate that the bottom currents transporting the sand at the time were essentially moving to the southeast and south-southeast, roughly paralleling a more northerly coastline. The bedding appears to represent a mass movement of sandwaves and large scale ripples along the bottom in those directions (Narayan, 1971; but see Middlemiss, 1975, p. 460, for discussion). The yellow sands are often extensively bioturbated by crustacean burrows, *Ophiomorpha*, indicating pauses in significant sand movement, allowing the organisms to proliferate and partly destroy primary bedding structures. The pebbles in the red carstone beds include quartzite, chert, siltstone and limonite of mixed derivation, but for most, from a northerly landmass. In constrast to the underlying Hythe Formation and Bargate Member beds, phosphatic pebbles are very sparse.

Stop 5. Newlands Corner [TQ 04314935].
This stop is included because it is close to and overlooks Albury, and in clear weather is a splendid viewpoint. At this location the Netley Heath Beds, of early Pleistocene age, rest uncomfortably on what is often said to be a Pliocene bevelled surface, at about 180 m O.D., of the Chalk. A Pleistocene marine fauna has been identified from the beds at the type-locality, 5 km to the east. When exposed they usually comprise ferruginous pebbly sands carrying flint and Lower Greensand cherts. Unfortunately, there are no good sections visible in the deposits at the Corner and traces have to be sought along pathways through the adjacent trees and bushes. Just below the crest is an excellent viewpoint looking southwards across into the centre of the Weald, with the depression formed by the Gault clay at the foot of the escarpment, and the distinct dip

Early Cretaceous Environments of the Weald

slopes of the Folkestone and Hythe Formations in the near and middle to far distance. As the Lower Greensand chert pebbles in the Netley Heath Beds have been derived probably from the rocks outcropping in front of you, it follows that the topography reflects erosion since at least early Pleistocene times.

Itinerary 10. Redhill - Bletchingley - Godstone
Alastair Ruffell

(Sheet 187, 1:50,000)

How to get there

The main outcrop of Lower Greensand in this area has traditionally been at the Nutfield fuller's earth quarries, east of Redhill, but now extraction has ceased.

In order to see the very variable succession in the Reigate - Redhill - Bletchingley area, seven separate locations, referable to the Nutfield type section, will be used: some are within the immediate vicinity of Redhill or Bletchingley, others cannot easily be reached by public transport. The suggested route is from Godstone to Redhill and back, in order follow something close to the stratigraphic succession from oldest to youngest.

Key literature

Casey (1961); Dines & Edmunds (1933); Gossling (1929, 1936); Kirkaldy (1947) Leighton (1895).

Summary of geological interest

A log of the succession mapped from quarry outcrops and boreholes by B.G.S. is given in Figure 29. (After Knox *et al.*, unpubl.). The locality numbers are indicated on this succession, which serves as a graphic summary of the main features to be seen, and allows the visitor to keep sight of their place in the stratigraphy.

Description

The Sandgate Formation in the Nutfield area of Redhill (Surrey) have been the principal source of fuller's earth (bentonite) since Roman times, and has accounted for around 65 per cent of total U.K. production since records began in 1854 (Highley, 1975). However, until the recent study, the only stratigraphical accounts have been those of Gossling (1929), Dines & Edmunds (1933) and Kirkaldy (1947). These accounts were based on shallow workings. Fuller's earth (bentonites) are a highly valued commodity, being used in drilling mud, as an industrial bonding agent, in waste disposal, confectionary,

Early Cretaceous Environments of the Weald

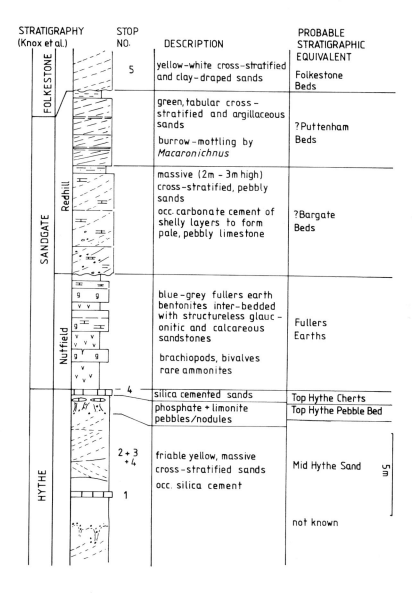

Figure 29. Summary stratigraphy of the Lower Greensand exposed in the Reigate - Redhill - Bletchingley area. Stop numbers are indicated adjacent to the stratigraphic section exposed.

Early Cretaceous Environments of the Weald

cosmetics, cat litter and in environmentally-friendly pesticide/fertiliser applications. The principal component of bentonite is calcium or sodium montmorillonite, an absorbent "swelling clay mineral".

Stratigraphic succession (mid-Aptian - early Albian), Redhill area

Folkestone Fm.		yellow - white - brown, cross-stratified sands, erosive base
Sandgate Fm.		
	Redhill "sands"	green - brown, mega-cross-stratified sands, rare calcareous cement, erosive base
	Nutfield "beds"	alternating fuller's earth (bentonites) & green fossiliferous sandstones
Hythe Fm.		Si or Ca cemented sands

Stop 1. Waterfall in Hythe Beds, Walkingstead Pond/Leigh Place [TQ 36215078].

How to get there

Walk down the road, opposite Fairall's Builder's Merchants (see below) and the waterfall is visible on the southern side.

Literature

There is very little literature of direct relevance. Gossling & Bull (1948) mapped in this area; a (temporary) adjacent location was mentioned by Batchelor & Ward (1990).

Description

This section is largely inaccessible for detailed study, as it is located around a waterfall opposite the pathway bridge. Nonetheless, it serves to demonstrate that the lower parts of the Hythe Formation in this area have some cemented horizons capable of forming resistant ledges during fluvial erosion. Such hardened beds are known to be calcareously cemented, with some chert (silica), and have given rise to a sub-division of the Hythe Formation known as the Lower Hythe Stone (Gossling, 1929). This contrasts with the soft Hythe

Early Cretaceous Environments of the Weald

Formation at the next two locations (below), and makes an interesting comparison with the hardened topmost beds of the Hythe in this area.

Stop 2. Taylor's Hill Pit, now Fairall's Builder Merchants, adjacent to Tilburstow Hill, Godstone [TQ35775065].

How to get there
    A wide gateway with sign adjoins the B2236 south of Goldstone: this allows room for parking, and most of the features of interest are visible from the entrance, with no need to enter the yard (a disused sandpit). Geologists wishing to study the sands in detail must seek verbal permission from Fairall's.

Key literature
    Gossling & Bull (1948); Kirkaldy (1947); Casey (1961); Ruffell (1992).

Description
The 20m high faces surrounding this yard display some of the best cross-stratified sands in the Hythe Formation. These beds are largely uncemented, in direct constrast to the Hythe above and below, hence their extraction for building sand, and generally poor exposure. Gossling (1929) recognised these sands in the middle of the Hythe Formation, dubbing them "Mid-Hythe Sands". Close inspection is not recommended here, but is possible at the next stop.

Stop 3. "Underhills" or Tilburstow Hill, south side [TQ347055023].

How to get there
    On South Park Lane, south of Godstone, a car-park is located to the south-east of Tilburstow Hill [TQ 34905012]. From here one may walk along the quiet South Park Lane westwards until a steep vegetated bank is reached. From here, climb up any one of the small tracks northwards until under the crest of Tilburstow Hill. "Underhill's" appears to be a local name, not used on modern maps, but appears in Dines & Edmunds (1933, figure 3). T. Bachelor (pers. comm., 1994) has indicated to the authors that the area known locally as "Underhills" is actually to the south of this stop.

Key literature
    Gossling & Bull (1948); Kirkaldy (1947).

Description
Here the 'Mid-Hythe Sands' discussed at the previous stop may be examined close-up. By contrast, the beds are found to be horizontally bedded or strongly bioturbated. White clay infills some vertical burrows that may penetrate 50 cm

Early Cretaceous Environments of the Weald

into the sand. The sand itself is medium-grained, with a few isolated pebbles. We may conclude from Taylor's Hill and from this location that the 'Mid-Hythe Sands' probably comprises variably cross stratified and strongly bioturbated sands: these may have no order vertically, but may pass laterally into one another, so caution must be exercised in correlation! The topmost section exposed is pebbly with some chert beds: Mr. T. Bachelor (pers. comm., 1994) suggests that these might be the upper beds of the Hythe Formation

Stop 4. Redhill Common, Redhill [TQ 273496].

How to get there
Redhill Common is not labelled on OS Sheet 187, but may be seen 1.5km southwest of Redhill railway station as a wooded area dissected by three pathways in a triangle.

Key literature
Dines & Edmunds (1933); Gossling (1929).

Description
The wooded bank at the eastern end of the common exposes patches of 'Mid-Hythe Sands', the 'Top Hythe Cherts' and fuller's earths. Each exposure is separate from the next, so the stratigraphic order is not clear unless known already, or seen at Bletchingley. This serves to demonstrate the excellent work of Leighton (1895), Gossling (1929) and Dines & Edmunds (1933) in interpreting the stratigraphy without the benefit of the deep workings for fuller's earth at Redhill. This is especially true when one considers that the beds wedge in and out in the Redhill area. For instance, at Redhill Common a bright crimson sand may be observed in the 'Mid-Hythe Sands'.

Stop 5. North Park Farm (sandpit), Godstone [TQ 3420 5178].

How to get there & summary
North Park Lane runs north from the A25 Godstone - Bletchingley Road at TQ 34275150. A public right of way crosses 200 m along this lane, to the west is a large and rambling sandpit. This is a new working in the Folkestone Formation sands that extends up to the footpath (and occasionally eats into it). The best view of the location is from the road as from the path one cannot see the whole working in perspective. Being a recently opened operation, there is no published literature: the stop is included here to provide a complete Lower Greensand stratigraphy.

Early Cretaceous Environments of the Weald

Itinerary 11. The eastern North Downs
Alastair Ruffell

(Sheet 188, 1:50,000)

Sevenoaks and Maidstone area - general geology
The geology of the Sevenoaks - Maidstone district is dominated by exposures of the Hythe Formation. There is one permanent Weald Clay exposure southwest of Pluckley and the Atherfield Clay is poorly exposed. The Sandgate Formation has a much reduced thickness in this area, a typical value being 4 m to 10 m, and this is frequently of red mudstone, thus making differentiation from drift difficult. A thicker Sandgate Formation succession, containing fuller's earths, was once quarried east of Maidstone. The Folkestone Formation sands are quarried north of Maidstone, but access is limited.

There are a number of Hythe Formation exposures in quarries in the Sevenoaks – Maidstone area, but the area is one of urbanisation and the state of many quarries changes rapidly. This may result in their use as landfill sites or for buildings. In the case of private housing this may temporarily preserve quarry faces, although access is not easy. Where industrial, privately owned buildings have been built, the quarries should be examined with the owners consent. Unfortunately, the turnover in ownership of large warehouses and industrial buildings is such that pre-excursion visits are the only method of securing definite entry. However, here are three of the "easiest" locations to visit, giving a representative view of the Hythe Formation of the area.

Stop 1. Dryhill [SU 498552].

How to get there
The Dryhill quarries are now a public park, signposted south from the A25 near its intersection with the A21 dual carriageway west of Sevenoaks. Although only 2.5 km from Sevenoaks town centre, walking to Dryhill is not easy as one has to use the minor road bridge at Salter's Heath (TQ 507548).

Key literature
Wright & Thomas (1946); Casey (1961); Worssam (1993).

Description
Dryhill once comprised a number of small quarries, now variably preserved throughout the park. In the lowland areas of the park small hollows with dipping "Rag (sandy, glauconitic limestone) & Hassock (clayey, glauconitic sand)" are common, whilst at the eastern limit of the roadway [TQ 49905526] a

Early Cretaceous Environments of the Weald

south – facing quarry face shows the Hythe Formation clearly, though with a lower dip than the beds in the old floor of the quarry. This difference is probably due to landslipping.

The variability in the degree of cementation of the bedded units is a matter for much speculation, some regarding it as a superficial Quaternary weathering effect, others as caused by cyclical changes in the chemical properties of the pore water soon after deposition (Middlemiss, 1975). Kentish Rag, both here and elsewhere in this area, was quarried extensively in the past from Roman times onwards, and constitutes the fabric of many old churches and other buildings in southeast England.

Stop 2. Ditton Industrial Estate, Maidstone [TQ 71775757 to 72345766].

How to get there
The disused quarry site at Ditton is now occupied by a wide variety of small and large industrial buildings. These are mainly of the corrugated warehouse type that fill each quarried area, leaving about 2 m of space for the geologist to pass between! Although an old public footpath traverses the estate, permission must be sought if close examination is intended. For the walking visitor Ditton makes an excellent excursion, the best route for which is to alight at Barming Station and follow one of the three east to west footpaths into Ditton, returning to East Malling or Barming (via the Royal British Legion Village). If driving, you must enter the area via Ditton housing estate to the north.

Key literature
Casey (1961); Ruffell (1992); Worssam (1993).

Description
The height of individual exposures of fossiliferous Rag & Hassock (Hythe Formation) is up to 2 m and they are usually weathered to a dusty appearance. Fossils such as sponges, brachiopods, bryozoa and belemnites are recorded (Ruffell, 1992), although extraction of such material is neither easy nor recommended.

Stop 3. Allington to Royal British Legion Industrial Estates and Quarry [TQ 74455798 to 73775785].

How to get there
The large quarries at the Royal British Legion Village and at Allington are occupied by industrial buildings (in the west) and still worked for occasional roadstone (in the east). Access is via the roundabout at TQ 73495765. The area is on open access, although large groups must

Early Cretaceous Environments of the Weald

gain permission from the relevant owners both in the west (Geest) and to the east (Allington Quarry). Access on foot is possible, although the public footpath emanating from the Allington Road at the railway bridge [TQ 74575748] now passes along the muddy quarry entrance road 100 m to the northeast (TQ 7468 5765). From there, the path leads directly into the quarry, where it terminates!

Key literature
Casey (1961); Worssam (1963); Ruffell (1992); Worssam (1993).

Description
The Hythe Formation dips at around 10° - 15° to the north, and lower beds, similar to those seen at Ditton, may be observed to the south and west of the quarry. To the east and north some exposures of the uppermost Hythe Formation, here termed the Boughton Member, may be seen. The stratigraphic position of the beds is known both from biostratigraphic analysis of the ammonite fauna (Casey, 1961) and by virtue of the lithostratigraphic succession. In the northeastern part of the Allington quarry, Sandgate Formation fuller's earths up to 1.2 m thick are known [around TQ 743580], and the Folkestone sands crop out (and are extracted) 2 km northwest at Aylesford [TQ 730592 to 724594]. The state of exposure at Allington is very variable: nonetheless fossil material may be picked up from loose blocks, although any large ammonites found by the quarrymen are usually sold to garden centres! Fossils include sponges, abundant wood fragments (sometimes replaced by pyrite) and brachiopods.

ACKNOWLEDGEMENTS

We have all benefited greatly in our studies of the Weald from the help and advice given by friends, colleagues and tutors over the years. Specifically, writing of the guide has been greatly aided through the help afforded by Perce Allen; Roger Bristow; Trevor Greensmith; Jake Hancock; Ed Jarzembowski; Frank Middlemiss and Bernard Worssam. AR would especially like to thank Brian Young for permission to publish details from his work on the Eastbourne sections and Trevor Bachelor who wrote a great deal of Itinerary 10 (Redhill).

Early Cretaceous Environments of the Weald

FURTHER READING

ALLEN, P. 1959. The Wealden environment: Anglo-Paris basin. *Phil. Trans. Roy. Soc., London*, **B242**, 283-346.

_____. 1960. Strandline pebbles in the mid-Hastings Beds and the geology of the London uplands. General features. Jurassic pebbles. *Proc. Geol. Ass.*, **71**, 156-168.

_____. 1961. Strandline pebbles in the mid-Hastings Beds and the geology of the London uplands: Carboniferous pebbles. *Proc. Geol. Ass.*, **72**, 271-285.

_____. 1967. Or igin of the Hastings facies in North-Western Europe. *Proc. Geol. Ass.*, **78**, 27-105.

_____. 1975. Wealden of the Weald: a new model. *Proc. Geol. Ass.*, **86**, 389-437.

_____. 1981. Pur suit of Wealden models. *Journ. Geol. Soc., London*, **138**, 375-405.

_____. 1989. Wealden research - ways ahead. *Proc. Geol. Ass.*, **100**, 529-564.

_____. & WIMBLEDON, W.A. 1991. Correlation of NW European Purbeck - Wealden (nonmarine Lower Cretaceous) as seen from the English type-areas. *Cret. Res.*, **12**, 511-526.

ANDERSON, I.D. 1986. The Gault Clay - Folkestone Beds junction in West Sussex, Southeast England. *Proc. Geol. Ass.*, **97**, 45-58.

ARKELL, W.J. 1939. Derived ammonites from the Lower Greensand of Surrey and their bearing on the tectonic history of the Hog's Back. *Proc. Geol. Ass.* **50**, 22-25.

BATCHELOR, T.B. & WARD, D. 1990. Fish remains from a temporary exposure of the Hythe Beds (Aptian - Lower Cretaceous) near Godstone, Surrey. *Mesozoic Research,* **2**, 181 - 203.

BRISTOW, C.R. 1981. *Geology of the country around Petersfield.* Int Rep. Inst. Geol. Sci.

BRISTOW, C.R. & MORTER, A.A. 1983. Field Meeting: A traverse of the Weald. 6th June 1982. *Proc. Geol. Ass.*, **94**, 377 - 381.

BRISTOW, C.R. & WYATT, R.J. 1983. Geological notes and local details details for 1:10,000 sheets TQ NW, NE, SE, & SW (Pulborough & Storrington). Keyworth: Inst.of Geol. Sci.

CASEY, R. 1961. The stratigraphical palaeontology of the Lower Greensand. *Palaeontology* **3**, 487 - 622.

Early Cretaceous Environments of the Weald

DINES, H.G. & EDMUNDS, F.H. 1929. *Geology of the country around Aldershot and Guildford.* Mem. Geol Surv. Eng. & Wales.

DINES, H.G. & EDMUNDS, F.H. 1933. *The geology of the country around Reigate and Dorking.* Memoir of the Geological Survey of England and Wales. Sheet 285, 204pp.

DODSON, M.H., REX, H.D., CASEY, R. & ALLEN, P. 1964. Glauconite dates from the Upper Jurassic and Lower Cretaceous. *Quart. J. Geol. Soc., Lond.*, **120,** 145-156.

GALE, A.S. 1989. Field meeting at Folkestone Warren, 29th November, 1987. *Proc. Geol. Ass.*, **100,** 73 - 82.

GALLOIS, R. W. & WORSSAM, B. C. 1993. *Geology of the country around Horsham.* Memoir of the British Geological Survey, No. 302, 130pp.

GOSSLING, F. 1929. The geology of the country around Reigate. *Proc. Geol. Ass.*, **40**, 197-259

GOSSLING, F. 1936. Field meeting at Oxted & Godstone. *Proc. Geol. Ass.,* **47,** 322-327.

GOSSLING, F. & BULL, A.J. 1948. The structure of Tilburstow Hill, Surrey. *Proc. Geol. Ass.*, 59, 131-139.

HALLAM, A. 1984. Continental humid and arid zones during the Jurassic and Cretaceous. *Palaeogeog., Palaeoclim., Palaeoecol.,* **47,** 195-223.

_____. 1986. Role of c limate in affecting late Jurassic and early Cretaceous sedimentation in the North Atlantic. *In:* SUMMERHAYES, C.P. & SHACKLETON, N.J. (eds) *North Atlantic Palaeogeography.* Geological Society of London, Special Publication, **21,** 277-281. or **26,** 251-256.

HESSELBO, S.P., COE, A.L. & JENKYNS, H.C. 1990. Recognition and documentation of depositional sequences at outcrop: an example from the Aptian & Albian on the eastern margin of the Wessex Basin. *Journ. Geol. Soc. Lond.,* **147,** 549-559.

HESSELBO, S.P. & ALLEN, P.A. 1991. Major erosion surfaces in the basal Wealden Beds, Lower Cretaceous, south Dorset. *Journ. Geol. Soc. Lon.,* **148,** 105-113.

HIGHLEY, D.E. 1975. The economic geology of the Weald. *Proc. Geol. Ass.,* **86,** 559-569.

HORNE, D. J. 1988. *Cretaceous Ostracoda of the Weald.* British Micropalaeontological Society Field Guide, No. 4, 42pp.

HOWITT, F. 1964. Stratigraphy and structure of the Purbeck inliers of Sussex (England). *Quart. J. Geol. Soc., Lond.,* **120,** 77-113.

Early Cretaceous Environments of the Weald

CASEY, R. 1963. The dawn of the Cretaceous period in Britain. *Bulletin of the south-eastern Union of Scientific Societies*, **117**, 1-15.

HUMPHRIES, D.W. 1964. The Stratigraphy of the Lower Greensand of the South-West Weald. *Proc. Geol. Ass.*, **75**, 39-59.

JARZEMBOWSKI, E. A. 1991. The Weald Clay of the Weald: report of 1988/89 field meetings. *Proc Geol Ass.*, **102**, 83-92.

JUKES-BROWNE, A.J. & W. HILL. 1900. *The Cretaceous rocks of Britain. 1. The Gault and Upper Greensand of England.* Mem. Geol. Surv., Gt. Br., 499pp.

KANTOROWICZ, J.D., 1990. Lateral and vertical variations in pedogenesis and other early diagenetic phenomena, Middle Jurassic Ravenscar Group, Yorkshire., *Proc. Yorks. Geol. Soc.*, **48**, 61-74.

KENNEDY, W.J. 1967. Field Meeting at Eastbourne, Sussex. Lower Chalk sedimentation. *Proc. Geol. Ass.*, **77**, 365-370.

KENNEDY, W.J., 1975. Morphology and genesis of nodular chalks and hardgrounds in the Upper Cretaceous of southern England. *Sedimentology*, **22**, 311-386.

KIRKALDY, J.F. 1932. The geology of the area around Hascombe, Surrey. *Proc. Geol. Ass.*, **43**, 127-151.

KIRKALDY, J.F. 1933. The Sandgate Beds of the Western Weald. *Proc. Geol. Ass.*, **34**, 270-311.

KIRKALDY, J.F. 1947. The work of the late Mr. Frank Gossling on the stratigraphy of the Lower Greensand between Brockham (Surrey) and Westerham (Kent). *Proc. Geol. Ass.*, **58**, 178-192.

KIRKALDY, J.F. 1976. (Revised by F.A. MIDDLEMISS, L.J. ALLCHIN & H.G. OWEN). *The Weald.* Geologists' Association Guide, **No. 29**, 27pp.

KIRKALDY, J.F. & WOOLDRIDGE, S.W. 1938. Notes on the geology of the country around Haslemere and Midhurst. *Proc. Geol. Ass.* **49**, 135-147.

KIRKALDY, J.F. & MIDDLEMISS, F.A. 1954. Field meeting in the Hindhead neighborhood. *Proc. Geol. Ass.*, **65**, 175-177

KNOWLES, L. & MIDDLEMISS, F.A. 1958. The Lower Greensand in the Hindhead area of Surrey and Hants. *Proc. Geol. Ass.*, **69**, 205-238.

KNOX, R.W.O'B., HIGHLEY, D. & RUFFELL, A. (in prep.) Stratigraphy of the Sandgate Beds (Lower Greensand; late Aptian) in the Nutfield area of southern England.

LAKE, R.D. 1975. The structure of the Weald. *Proc. Geol. Ass.*, **86**, 549-557.

LAKE, R.D. & SHEPHARD-THORN, E.R. 1985. The stratigraphy and geological structure of the Hog's Back, Surrey, and adjoining areas. *Proc. Geol. Ass.*, **96,** 7-21.

LEIGHTON, T. 1895. The Lower Greensand above the Atherfield Clay of East Surrey. *Quart. J. Geol. Soc., Lond.* **51,** 101-124.

MIDDLEMISS, F.A. 1961a. A fauna from the Puttenham Beds (Lower Greensand) of Hampshire. *Proc. Geol. Ass.*, **72,** 455 - 459.

MIDDLEMISS, F.A., 1961b. Field meeting in the western end of the Weald. *Proc. Geol. Ass.*, **73,** 125 - 129.

MIDDLEMISS, F.A. 1962a. Brachiopod ecology and Lower Greensand palaeogeography. *Palaeontology,* **5,** 253 - 267.

MIDDLEMISS, F.A., 1962b. Vermiform burrows and rates of sedimentation in the Lower Greensand. *Geol. Mag.*, **99,** 33 - 40.

MIDDLEMISS, F.A., 1975. Studies in the sedimentation of the Lower Greensand of the Weald, 1875-1975: a review and commentary. *Proc. Geol. Ass.*, **86,** 457-473.

MIDDLEMISS, F.A. 1978. The cherts in the Hythe Beds (Lower Cretaceous) of S.E. England. *Proc. Geol. Ass.*, **89,** 283-298.

NARAYAN, J. 1971. Sedimentary structures in the Lower Greensand of the Weald, England, and Bas-Boulonnais, France. *Sedim. Geol.*, **6,** 73-109.

OWEN, H.G. 1971. Middle Albian stratigraphy in the Anglo-Paris Basin. *Bull. Br. Mus. Nat. Hist. (Geol).* Suppl. **8.**

OWEN, H.G. 1976. The stratigraphy of the Gault and Upper Greensand of the Weald. *Proc. Geol. Ass.*, **86,** 475-98.

OWEN, H.G. 1988. The ammonite zonal sequence and ammonite taxonomy in the *Douvilleiceras mammillatum* Superzone (Lower Albian) in Europe. *Bull. Brit. Mus. Nat. Hist. (Geol.),* **44,** 177 - 231.

PADGHAM, R.C. 1972. Field meeting to the Folkestone Beds (Lower Greensand) of West Surrey. *Proc. Geol. Ass.*, **83,** 355-359.

PRICE, F.G.H. 1874. On the Gault of Folkestone. *Quart. J. Geol. Soc., Lond.,* **30,** 342 - 366.

RAWSON, P.F. & RILEY, L.A. 1982. Latest Jurassic - Early Cretaceous events and the "late Cimmerian unconformity" in North Sea area. *Bull. Amer. Assoc. Petrol. Geologists,* **66,** 2628-2648.

RAWSON, P.F. 1992. Early Cretaceous. In: J.C.W. Cope, J.K. Ingham and P.F. Rawson (Editors), *Atlas of Palaeogeography and Lithofacies.* Geol. Soc. Lond., Memoir 13, 131-135.

REID, C. 1903. *The geology of the country near Chichester.* Mem. Geol. Surv. Gt. Brit.

RICHARDSON, J.A. 1947. Chert formation in the Bargate Beds of the

Churt neighbourhood, Surrey. *Proc. Geol. Ass.,* **58**, 161-77.
ROSS, A.J. & COOK, E. 1995. The stratigraphy and palaeontology of the Upper Weald Clay (Barremian) at Smokejacks Brickworks, Ockley, Surrey, England. *Cretaceous Research,* **16**, 705-16
RUFFELL, A.H., 1990, The mineralogy and petrography of the Sulphur Band phosphates (Aptian-Albian), at Folkestone, Kent. *Proc. Geol. Ass.,* **101**, 79-84.
RUFFELL, A.H. 1992. Correlation of the Hythe Beds Formation (Lower Greensand Group: early - mid-Aptian, southern England. *Proc. Geol. Ass.,* **103**, 273 - 291.
SELLWOOD, B.W., SCOTT, J. & LUNN, G. 1986. Mesozoic basin evolution in southern England. *Proc. Geol. Ass.,* **97**, 259-289.
SHEPHERD, W.B. 1934. Some observations on the Folkestone Beds around Farnham. *Proc. Geol. Ass.,* **45**, 436.
SLADEN, C.P. 1983. Trends in Early Cretaceous clay mineralogy in NW Europe. *Zitteliana,* **10**, 349-357.
_____. & B ATTEN, D.J. 1984. Source-area environments of the Late Jurassic and Early Cretaceous sediments in Southeast England. *Proc. Geol. Ass.,* **95**, 149-163.
SMART, J.G.O., BISSON, G. & WORSSAM, B. 1966. *Geology of the country around Canterbury and Folkestone.* Mem. Geol. Surv, Gt. Br. 337pp.
STEWART, D.J. 1981. A meander-belt sandstone of the Lower Cretaceous of southern England. *Sedimentology,* **28**, 1-20.
STEWART, D.J., 1983. Possible suspended-load channel deposits from the Wealden Group (Lower Cretaceous) of southern England. *Spec. Publ. Int. Ass. Sediment.,* **3**, 369-384.
STONELEY, R. 1982. The structural development of the Wessex Basin. *Journ. Geol., Soc., Lond.,* **139**, 545-552.
TAYLOR, K.G. 1990. Berthierine from the non-marine Wealden (early Cretaceous) sediments of south-east England. *Clay Minerals,* **25**, 391-399.
_____.1991. Phospha tic concretions in the Wealden of South-East England. *Proc. Geol. Assoc.,* **102**, 67-70.
_____. 1992. Non-mar ine oolitic ironstones in the Lower Cretaceous Wealden sediments of southeast England. *Geol. Mag.,* **129**, 349-358.
THURRELL, R.G., WORSSAM, B.C. & EDMONDS, E.A. 1968. *Geology of the country around Haslemere.* Mem. Geol. Surv. Gt.Br. 169pp.
TOPLEY, W. 1875. *The Geology of the Weald.* Memoirs of the

Geological Survey, United Kingdom. 503pp.
WACH, G.D. & RUFFELL, A.H. 1991. *Sedimentology & sequence stratigraphy of a Lower Cretaceous tide and storm - dominated clastic succession, Isle of Wight and S.E.England.* Field Guide 4, International Sedimentological Congress, Nottingham. 100pp.
WHITE, H.J.O. 1924. *The geology of the country near Brighton and Worthing.* Mem. Geol. Surv. Eng. & Wales.
WHITTAKER, A. (ed.) 1985. *Atlas of Onshore Sedimentary Basins in England and Wales: Post-Carboniferous Tectonics and Stratigraphy.* Blackie, London & Glasgow.
WOOD, G.V. 1957. The heavy mineral suites of the Lower Greensand of the Western Weald. *Proc. Geol. Ass.* **67,** 124-137.
WOODWARD, A.S. 1911. On some mammalian teeth from the Wealden of Hastings. *Quart. J. Geol. Soc. Lond.,* **67,** 278-281.
WOOLDRIDGE, S.W. 1947. Haslemere and Midhurst: Whitsun Field Meeting to the central Weald. *Proc. Geol. Ass.,* **58,** 73-76.
WORSSAM, B.C. 1963. *Geology of the country around Maidstone.* Mem. Geol. Surv. Eng. & Wales.
WORSSAM, B.C., 1964, Iron ore workings in the Weald Clay of the Western Weald. *Proc. Geol. Ass.,* **75,** 529-546.
WORSSAM, B.C. 1978. *The stratigraphy of the Weald Clay.* Rep. Inst. Geol. Sci., **78/11,** 23pp.
WORSSAM, 1993. Correlation of the Hythe Beds Formation (Lower Greensand Group: early - mid-Aptian, southern England: discussion, with Reply by A.RUFFELL. *Proc. Geol. Ass.,* **104,** 301 - 307.
WRIGHT, C.W. & THOMAS, H.D. 1946. Notes on the geology of the country around Sevenoaks, Kent. *Proc. Geol. Ass.,* **57,** 315-321.
YOUNG, B. 1978. The Upper Greensand of Eastbourne, Sussex. *Guide Book to the Sixth International Clay Conference, Oxford.* pp. 48-52
YOUNG, B. & MORGAN, D.J. 1981. The Aptian Lower Greensand fuller's earth beds of Bognor Common, West Sussex. *Proc. Geol. Ass.,* **92,** 33-37.
YOUNG, B. & LAKE, R.D. 1988. *Geology of the country around Brighton & Worthing.* Mem. Br. Geol. Surv. Nos. 318 & 333.